Science and Technology Concepts for Middle Schools™

Human Body Systems

**Student
Guide
and
Source
Book**

NATIONAL SCIENCE RESOURCES CENTER

The National Science Resources Center (NSRC) is operated by the Smithsonian Institution and the National Academies to improve the teaching of science in the nation's schools. The NSRC disseminates information about exemplary teaching resources, develops curriculum materials, and conducts outreach programs of leadership development and technical assistance to help school districts implement inquiry-centered science programs.

SMITHSONIAN INSTITUTION

The Smithsonian Institution was created by act of Congress in 1846 "for the increase and diffusion of knowledge. . . ." This independent federal establishment is the world's largest museum complex and is responsible for public and scholarly activities, exhibitions, and research projects nationwide and overseas. Among the objectives of the Smithsonian is the application of its unique resources to enhance elementary and secondary education.

THE NATIONAL ACADEMIES

The National Academies are nonprofit organizations that provide independent advice to the nation on matters of science, technology, and medicine. The National Academies consist of four organizations: the National Academy of Sciences, the National Academy of Engineering, the Institute of Medicine, and the National Research Council. The National Academy of Sciences was created in 1863 by a congressional charter. Under this charter, the National Research Council was established in 1916, the National Academy of Engineering in 1964, and the Institute of Medicine in 1970.

STC/MS PROJECT SPONSORS

National Science Foundation
Bristol-Myers Squibb Foundation
Dow Chemical Company
DuPont Company
Hewlett-Packard Company
The Robert Wood Johnson Foundation
Carolina Biological Supply Company

Science and Technology Concepts for Middle Schools™

Human Body Systems

Student Guide and Source Book

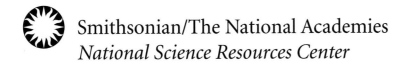

Smithsonian/The National Academies
National Science Resources Center

Published by Carolina Biological Supply Company
Burlington, North Carolina

NOTICE This material is based upon work supported by the National Science Foundation under Grant No. ESI-9618091. Any opinions, findings, and conclusions or recommendations expressed in this material are those of the authors and do not necessarily reflect views of the National Science Foundation, the Smithsonian Institution, or the National Academies.

This project was supported, in part,
by the
National Science Foundation
Opinions expressed are those of the authors
and not necessarily those of the Foundation

ISBN 0-89278-853-4

Published by Carolina Biological Supply Company, 2700 York Road, Burlington, NC 27215.
Call toll free 1-800-334-5551.

Cover design and illustration by Max-Karl Winkler; cover photo by Renée Bouchard/NSRC.
Printed in the United States of America

CB787700003
♻ Printed on recycled paper.

Human Body Systems

MODULE DEVELOPMENT STAFF

Developer/Writer
Henry Milne

Science Advisor
Jayne Hart, Professor of Biology,
George Mason University

Editor
Linda Harteker

Contributing Writers
Linda Harteker
Patrick Zickler

Illustrators
Taina Litwak
Max-Karl Winkler

Photographic Research
Matthew Bailey
Carolyn Hanson
PhotoAssist, Inc

Design Consultation
Isely &/or Clark Design

STC/MS Project Staff

Principal Investigators
Douglas Lapp, Executive Director, NSRC
Sally Goetz Shuler, Deputy Director, NSRC

Project Director
Kitty Lou Smith

Publications Director
Heather Dittbrenner

Managing Editor
Dorothy Sawicki

Senior Editor
Linda Harteker

Illustration Coordinator
Max-Karl Winkler

Photo Editor
Janice Campion

Graphic Designer
Heidi M. Kupke

Administrative Officer
Gail Thomas

Program Assistants
Matthew Bailey
Carolyn Hanson

Publications Assistant
Famin Ahmed

STC/MS Project Advisors

Judy Barille, Chemistry Teacher, Fairfax County, Virginia, Public Schools

Steve Christiansen, Science Instructional Specialist, Montgomery County, Maryland, Public Schools

John Collette, Director of Scientific Affairs (retired), DuPont Company

Cristine Creange, Biology Teacher, Fairfax County, Virginia, Public Schools

Robert DeHaan, Professor of Physiology, Emory University Medical School

Stan Doore, Meteorologist (retired), National Oceanic and Atmospheric Administration, National Weather Service

Ann Dorr, Earth Science Teacher (retired), Fairfax County, Virginia, Public Schools; Board Member, Minerals Information Institute

Yvonne Forsberg, Physiologist, Howard Hughes Medical Center

John Gastineau, Physics Consultant, Vernier Corporation

Patricia Hagan, Science Project Specialist, Montgomery County, Maryland, Public Schools

Alfred Hall, Staff Associate, Eisenhower Regional Consortium at Appalachian Educational Laboratory

Connie Hames, Geology Teacher, Stafford County, Virginia, Public Schools

Jayne Hart, Professor of Biology, George Mason University

Michelle Kipke, Director, Forum on Adolescence, Institute of Medicine

John Layman, Professor Emeritus of Physics, University of Maryland

Thomas Liao, Professor of Engineering, State University of New York at Stony Brook

Ian MacGregor, Senior Science Advisor, Geoscience Education, National Science Foundation

Ed Mathews, Physical Science Teacher, Fairfax County, Virginia, Public Schools

Ted Maxwell, Geomorphologist, National Air and Space Museum, Smithsonian Institution

Tom O'Haver, Professor of Chemistry/Science Education, University of Maryland

Robert Ridky, Professor of Geology, University of Maryland

Mary Alice Robinson, Science Teacher, Stafford County, Virginia, Public Schools

Bob Ryan, Chief Meteorologist, WRC Channel 4, Washington, D.C.

Michael John Tinnesand, Head, K-12 Science, American Chemical Society

Grant Woodwell, Professor of Geology, Mary Washington College

Thomas Wright, Geologist (emeritus), U.S. Geological Survey; Museum of Natural History, Smithsonian Institution

Foreword

Community leaders and state and local school officials across the country are recognizing the need to implement science education programs consistent with the National Science Education Standards to attain the important national goal of scientific literacy for all students in the 21st century. The Standards present a bold vision of science education. They identify what students at various levels should know and be able to do. They also emphasize the importance of transforming the science curriculum to enable students to engage actively in scientific inquiry as a way to develop conceptual understanding as well as problem-solving skills.

The development of effective standards-based, inquiry-centered curriculum materials is a key step in achieving scientific literacy. The National Science Resources Center (NSRC) has responded to this challenge through the Science and Technology Concepts for Middle Schools (STC/MS) program. Prior to the development of these materials, there were very few science curriculum resources for middle school students that embody scientific inquiry and hands-on learning. With the publication of the STC/MS modules, schools will have a rich set of curriculum resources to fill this need.

Since its founding in 1985, the NSRC has made many significant contributions to the goal of achieving scientific literacy for all students. In addition to developing the Science and Technology for Children (STC) program—an inquiry-centered science curriculum for grades K through 6—the NSRC has been active in disseminating information on science teaching resources, in preparing school district leaders to spearhead science education reform, and in providing technical assistance to school districts. These programs have had a significant impact on science education throughout the country.

The transformation of science education is a challenging task that will continue to require the kind of strategic thinking and insistence on excellence that the NSRC has demonstrated in all of its curriculum development and outreach programs. Its sponsoring organizations, the Smithsonian Institution and the National Academies, take great pride in the publication of this exciting new science program for middle schools.

J. DENNIS O'CONNOR
Provost
Smithsonian Institution

BRUCE M. ALBERTS
President
National Academy of Sciences

Preface

The National Science Resources Center (NSRC) is dedicated to the development, dissemination, and implementation of innovative, hands-on science education programs. After the completion of the Science and Technology for Children (STC) program for elementary schools, the NSRC launched in 1997 the Science and Technology Concepts for Middle Schools (STC/MS) project. The STC/MS program is designed to meet the challenge of the National Science Education Standards to place scientific inquiry at the core of science education programs.

The STC/MS program, which consists of eight instructional modules, is designed to provide students with stimulating experiences in the life, earth, and physical sciences and in technology while simultaneously developing their critical-thinking and problem-solving skills. The NSRC believes that the way to do this is to engage students in scientific inquiry. The National Science Education Standards state: "Students in grades 5–8 should be provided opportunities to engage in full and partial inquiries. . . . With an appropriate curriculum and adequate instruction, middle school students can develop the skills of investigation and the understanding that scientific inquiry is guided by knowledge, observations, ideas, and questions."

Bruce Alberts, president of the National Academy of Sciences, reflects on the importance of teaching science through inquiry in the National Academies' publication *Inquiry and the National Science Education Standards: A Guide for Teaching and Learning*:

> Teaching science through inquiry allows students to conceptualize a question and then seek possible explanations that respond to that question. . . . Inquiry is in part a state of mind—that of inquisitiveness. Most young children are naturally curious. They care enough to ask "why" and "how" questions. But if adults dismiss their incessant questions as silly and uninteresting, students can lose this gift of curiosity. Visit any second-grade classroom and you will generally find a class bursting with energy and excitement, where children are eager to make new observations and try to figure things out. What a contrast with many eighth-grade classes, where the students so often seem bored and disengaged from learning and from school!

The STC/MS modules developed by the NSRC keep inquiry at the center of the learning process to encourage student curiosity—even in eighth graders. And the materials are unique in a number of other ways. The NSRC has developed each module using a rigorous research and development process. The STC/MS activities have been developed through repetitive cycles of classroom testing, review, and improvement. This research and development process has included both trial teaching and field-testing nationwide, as well as the active involvement of many scientific experts from universities, museums, government agencies, and industry. The NSRC has also designed special apparatus for many of the activities and tested each piece of equipment to perfect the design. After field testing, the STC/MS developers continued to revise the materials and apparatus, based on feedback from students, teachers, and experts.

The research and development process of the STC/MS curriculum involved a very productive collaboration of master teachers and scientists. Beginning with the conceptualization of each module, scientists have been involved, reviewing the conceptual structure and contributing to the learning activities in each module. Expert middle school science teachers have also worked with NSRC developers—master teachers themselves—to assess how students respond to the activities and to suggest ways to improve them. This collaboration, involving both scientists and expert teachers, has ensured that the learning activities in each module reflect current scientific thinking and are effective in the classroom. The involvement of such experts from the beginning has sparked creativity in the module development process and has added originality to every lesson.

Because this research and development process is time-consuming and labor-intensive—and therefore expensive—it is not surprising that few traditional science textbooks have been developed this way. The NSRC has received major support from the National Science Foundation and from many corporate and philanthropic foundations to develop the STC/MS program.

Thus, three unique factors—keeping inquiry at the center of each lesson, following a rigorous research and development process, and engaging the active collaboration of scientists and expert teachers—have characterized the development of the STC/MS program. This process has enabled the NSRC to produce a focused, inquiry-centered curriculum for middle schools that actively engages students in learning new science and technology concepts, while building critical-thinking and problem-solving skills that will be useful to them throughout their adult lives.

The NSRC is grateful to Kitty Lou Smith, STC/MS Project Director, for her tireless efforts and creative leadership of this project. Working in partnership with Managing Editor Dorothy Sawicki, Dr. Smith has guided her staff through all the phases of the arduous research and development process that has led to the publication of the STC/MS program modules.

We would also like to thank the NSRC's parent institutions, the Smithsonian Institution and the National Academies, for their vision and support in helping the NSRC to undertake this project. We look forward to hearing from teachers regarding their classroom experience with the STC/MS modules, together with any suggestions they may have for improvements.*

DOUGLAS M. LAPP
Executive Director
National Science Resources Center

*Suggestions and feedback can be sent by e-mail to: stcms@nas.edu, or mailed to: STC/MS Program, National Science Resources Center, Smithsonian Institution, Washington DC 20560-0403.

Acknowledgments

The National Science Resources Center gratefully acknowledges the following individuals and school systems for their assistance with the national field-testing of *Human Body Systems*:

The Einstein Project, Green Bay, Wisconsin

Site Coordinator: Sue Theno, Director of the Einstein Project

Green Bay Area Public School District

Dale M. De Villers, Teacher, Parkview Middle School

Dan Doersch, Teacher, Seymour Middle School

Mary Conard, Teacher, DePere Middle School

Montgomery County Public Schools, Montgomery County, Maryland

Secondary Science Coordinator: Gerry Consuegra

Site Coordinator: Patricia Hagan, Science Project Specialist

Pam Fountain, Science Department Chairperson/Teacher, Tilden Middle School

Karen Rabin, Principal, Tilden Middle School

Barry Vidrick, Life Science Teacher, Tilden Middle School

Bozeman Public School District, Bozeman, Montana

Site Coordinator: Myra Miller, Keystone Project Director

Wendy Pierce, Teacher, Chief Joseph Middle School

Doug Whitmer, Teacher, Sacajawea Middle School

Chris Ottey, Teacher, Sacajawea Middle School

West Windsor-Plainsboro Regional School District, New Jersey

Site Coordinator: Barbara Braverman, Math/Science Supervisor of Curriculum and Instruction, West Windsor-Plainsboro Middle School

Leslie Bush, Teacher, West Windsor-Plainsboro Middle School

Shirley Allan, Teacher, Thomas Grover Middle School

Miriam A. Robin, Supervisor of Science, Grades 6-8, Community Middle School

Strong School District, Maine

Site Coordinator: Jeanne Tucker, Mt. Abram High School

Verna Holman, Teacher, Stratton Elementary School

Gayle Petrie, Teacher, Kingfield Elementary School

Misty Young, Teacher, Strong Elementary School

Atlanta Public School District, Georgia

Site Coordinator: Stephanie May, Program Coordinator, Health Science Guides, Emory University

Yolanda Chaplin, Teacher, Marshall Middle School

The NSRC also thanks the following individuals from Carolina Biological Supply Company for their contribution to the development of this module—

Helen Kreuzer, Director of Product Development

Bobby Mize, Instructional Materials Manager

Erin Krellwitz, Product Developer

Ben Harris, Product Developer

Jennifer Manske, Publications Manager
Jonathan Shectman, Editor
Anita Wilson, Designer

Finally, the NSRC appreciates the contribution of its STC/MS project evaluation consultants:

Program Evaluation Research Group (PERG), Lesley College—

Sabra Lee, Researcher, PERG
George Hein, Director (retired), PERG

Center for the Study of Testing, Evaluation, and Education Policy (CSTEEP), Boston College—

Joseph Pedulla, Director, CSTEEP
Maryellen Harmon, Director (retired), CSTEEP

Contents

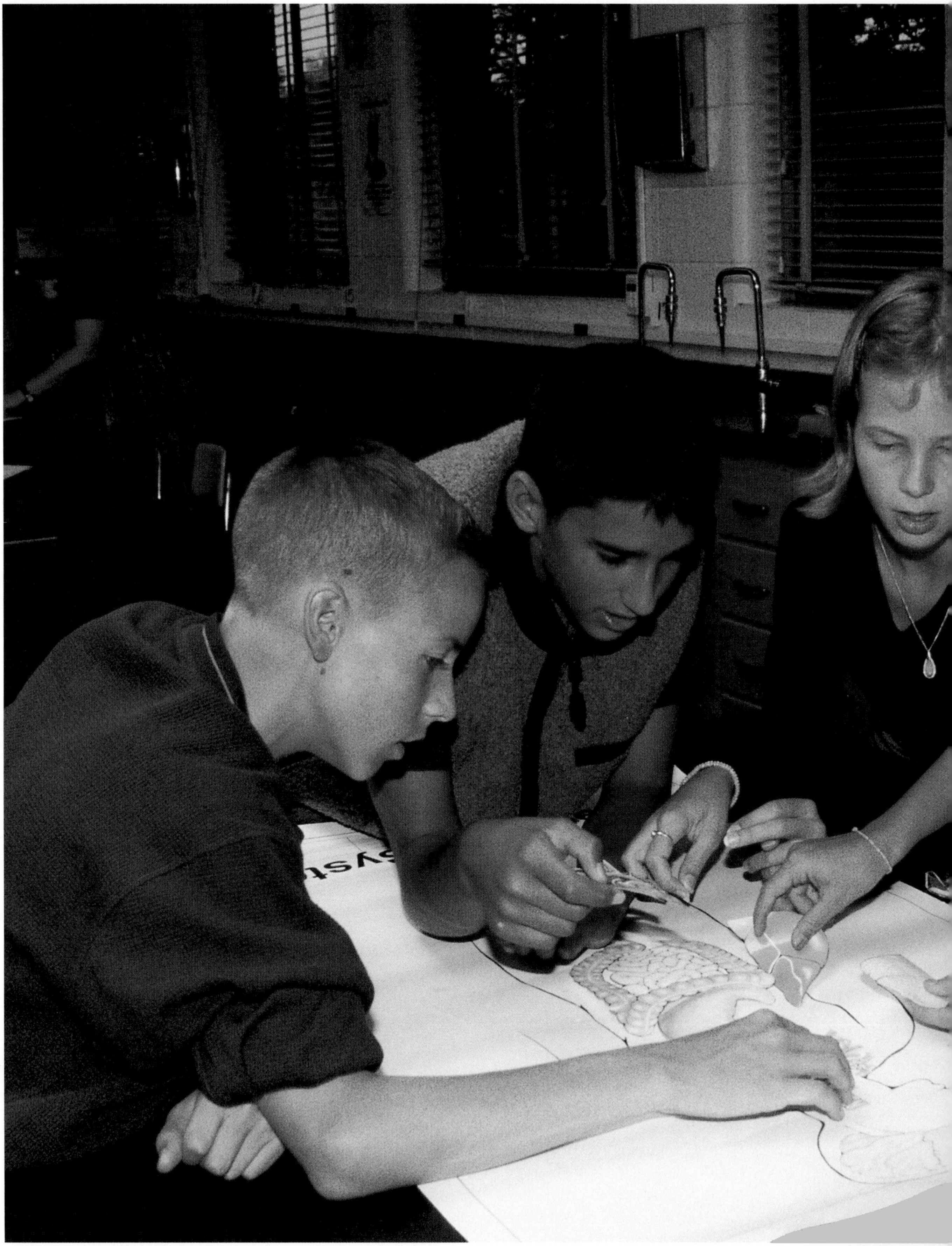

Human Body Systems— A Preassessment

© DAN HABIB/IMPACT VISUALS/PNI

How many healthy body systems do you need to play a hard-hitting game of volleyball? You may be surprised!

INTRODUCTION

How much do you know about your body? Do you know how your body digests food? Can you explain how your heart works? Why does your heart beat faster when you are exercising?

In this module, you will investigate four major human body systems and explore the answers to questions such as these. You will begin your exploration by thinking about what body systems and organs are. Then you'll discuss your ideas with your group. Next, your group will create a poster that illustrates some of the major organs of the body. Your group will use this poster throughout the module. You'll add new things to it and revise it when necessary.

OBJECTIVES FOR THIS LESSON

Construct definitions of the words "organ" and "body system."

Identify what you think are the major systems and organs of the human body.

Assemble a poster showing the positions and names of some of the major organs of the body.

GALEN—THE GLADIATORS' DOCTOR

CORBIS/BETTMANN

Galen

You might say that Galen, a Greek doctor who lived nearly 2000 years ago, was one of the first specialists in sports medicine. Galen was the official doctor for the gladiators in the city of Pergamum. Gladiators were professional fighters who waged bloody sword battles. Their opponents included wild beasts as well as other gladiators! Gladiatorial combat was a popular spectacle for people who lived at the time of the Roman Empire.

When treating his patients, Galen studied the gladiators' bones, muscles, and body systems. He made many discoveries. For example, he identified 300 muscles. He also proposed a system that described how the arteries carry blood.

Although not all Galen's discoveries were correct, they prepared the way for the investigations of later scientists such as Leonardo da Vinci, who lived in Italy during the 16th century.

MATERIALS FOR LESSON 1

For your group
1 human body systems poster
1 large sheet with illustrations of human organs
1 tote tray
4 pairs of scissors
1 roll of clear tape
1 30-centimeter (cm) ruler
1 black marker
1 large paper clip
1 eraser

Getting Started

1. What do you think is meant by a "body system"? What is an "organ"? What are the differences between body systems and organs? Discuss these questions with your group. Write your ideas on the first page of your science notebook.

2. Turn to the next blank sheet in your notebook. Divide it into two vertical columns. Label the left column "Organs" and the right column "Body Systems." With your group, brainstorm a list of the organs of the body. Write the names of the organs in the left column. Then brainstorm a list of human body systems. Enter the names of those systems in the right column.

3. Share your answers with the rest of the class.

Inquiry 1.1
Human Body Mapping

PROCEDURE

1. Have two members of your group pick up a tote tray, a human body systems poster, and a copy of the sheet with the illustrations of the organs.

2. Spread out the poster for everyone in the group to see. Have each member of the group write his or her name on the poster in one of the spaces provided.

3. Personalize your group's poster by agreeing on a name for it. Have one member of the group print the name in the upper left corner.

4. Have one person in your group make a *rough* cut around each of the organs on the sheet. Divide the organs so that each person in the group has two or three organs. Using the scissors, neatly cut out each organ, as shown in Figure 1.1.

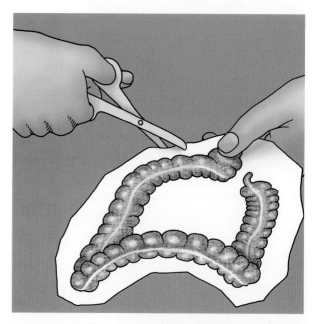

Figure 1.1 *Cutting out an organ for the human body systems poster*

5. After the organs are cut out, work with your group to agree on the proper name for each organ.

6. With your group, decide where each organ should go on the human body outline. You may place some organs on top of others.

7. After your group has agreed on where each organ goes, tape the organs to the poster. Fasten each cutout to the poster by doubling a piece of clear tape, sticking it on the back of the organ, and pressing the organ against the poster. The technique for folding the tape is shown in Figure 1.2.

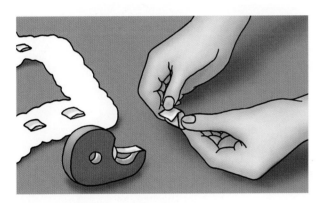

Figure 1.2 *How to fold the tape*

8. Using the ruler and marker, draw a straight line outward from each organ into the space to its left or right, as shown in Figure 1.3. Print the name of the organ neatly at the end of the line.

9. Follow your teacher's instructions for cleanup.

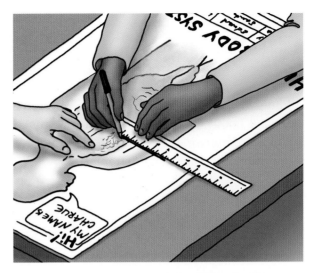

Figure 1.3 *Draw a line to the left or right of each organ.*

REFLECTING ON WHAT YOU'VE DONE

1. With your group, take a second look at the information you recorded in your science notebook at the beginning of this lesson. Revise the information if necessary.

2. Use the information you recorded, along with what you did in this inquiry, to develop definitions of "organ" and "body system." Write the definitions in your science notebook.

3. Share your definitions with the class.

4. This module is titled *Human Body Systems.* What other types of systems have you heard of? What do these systems seem to have in common? What might these other systems have in common with human body systems? Discuss your ideas with the class.

Humans—The Problem-Solving Animals

All humans are animals, but not all animals are human. What's the difference?

How We're Alike
Like most other animals, humans can move about. We're not rooted to one spot, like most plants are. Like some animals, we have a spine. Like apes, we have brains that are relatively large compared with the size of our bodies. Like apes, we have thumbs as well as fingers. This means we can make and use tools.

Humans and some animals share other traits that may be less apparent. For example, both humans and some animals are curious about their environments. Both have the ability to learn behaviors.

How We're Different
What is the difference between humans and other animals? One of the most important differences is that humans are very effective problem solvers. Humans can tackle abstract questions, as well as concrete problems.

Solving problems takes more than curiosity and intelligence. It often takes cooperation—the ability to work together. Humans are especially good at this, and in science, cooperation produces many rewards.

Scientists are great problem solvers. Although some important scientific

Dr. James Watson (left) and Dr. Francis Crick showed that teamwork pays off. In 1953, they discovered the structure of the DNA molecule. DNA stands for "deoxyribonucleic acid." Discovering the structure of DNA was an important advance in genetics. The two received the Nobel Prize for their work in 1962. Dr. Watson was only 25 years old when he and Dr. Crick made their important discovery.

discoveries have been made by geniuses working alone in their labs, most discoveries are the result of teamwork.

Science and Teamwork

Today, few scientists work alone. They work as members of research teams. Each scientist may have a slightly different area of expertise. The scientists talk, make inquiries, rethink their ideas, and reach a conclusion. They publish their results. Teams of scientists in other laboratories throughout the world study the new findings and try to duplicate and build on them.

You're a Scientist, Too

In this module, you will work with your group to explore human body systems. The inquiries you will do are based on the work of many famous scientists.

But don't think that all the discovery is done! Scientists still have a lot to learn. So be curious. Work cooperatively. Apply your special problem-solving skills.

Who knows what you might discover? ☐

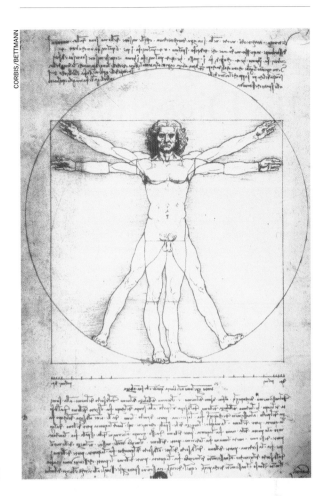

CORBIS/BETTMANN

QUESTIONS

1. What are some similarities and differences between humans and other animals?
2. What important scientific or technological breakthroughs have occurred in your lifetime? What scientists were responsible for them?
3. Why do you think that teamwork is important in science?

PERFECT PROPORTIONS

In the 16th century, scientists were looking for signs of harmony between the natural world and the world of mathematics. This drawing shows the perfect proportions of the human form. It aligns with a circle as well as a square.

Moving Through the Digestive Tract

How can a snake swallow and digest an egg? Humans aren't capable of feats like this. Our jaws aren't flexible enough, for one thing. The human digestive system, however, moves food along with the same kind of muscular contractions that the snake uses.

INTRODUCTION

In Lesson 1, you discussed what you know about human body systems and the organs of the body. For the next several lessons, you will focus on one body system—the digestive system. To begin, your group will perform an inquiry that will help you understand how food moves through the digestive tract. It's a long trip. Even though you are probably about 1.6 meters tall, your digestive tract is approximately 8 to 10 meters long!

After the introduction to the digestive system in this lesson, you'll be ready to take a closer look at what happens in each organ of the digestive system. That investigation will begin in Lesson 4.

OBJECTIVES FOR THIS LESSON

Discuss the purpose of the digestive processes.

Build a model of the digestive tract.

Investigate how food moves through the digestive tract.

Explore the function of mucus in the digestive tract.

Getting Started

1. Write the following questions in your science notebook. Leave several lines of space between each question. Then answer the questions.

   A. How do you think food moves through the digestive tract?

   B. What do you think happens to food as it moves through the digestive tract?

   C. Why do you think the digestive tract gets narrower at some places? (See Figure 2.1.)

2. Discuss your answers with your group. Agree on a single answer to each question. Select a spokesperson to present your group's answers to the class.

3. Listen as each group shares its answers. Add any new ideas that you think are important in your science notebook.

MATERIALS FOR
LESSON 2

For your group
1 tote tray
1 long piece of plastic tubing
1 pair of scissors
1 black marker
1 tennis ball (soaked in oil and stored in a fold-top plastic bag)
1 tape measure
1 large plastic storage bag
1 small plastic bag

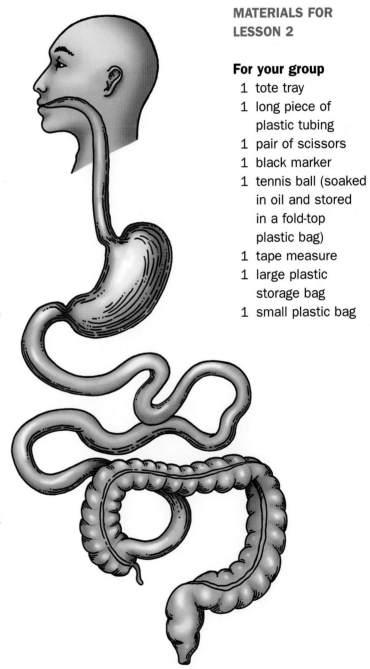

Figure 2.1 *A simplified drawing of the human digestive tract*

Inquiry 2.1
Moving Right Along

PROCEDURE

1. Select one person to pick up the materials for your group. Listen as your teacher reviews the Procedure for the inquiry.

2. Place the plastic tubing on a flat surface. Unroll about 40 cm of the tube.

3. Using Figure 2.2 and Table 2.1 for guidance, measure and mark on the tubing the approximate length of each organ of the digestive tract. Start at the opening of the tube, which represents the opening of the mouth. The mouth is about 11 cm long; therefore, you should place your first mark 11 cm from the opening of the tube. Using that mark as a starting point, then measure 25 cm more, which will represent the esophagus. Continue measuring and marking until you reach the anus, the opening at the end of the rectum. Double-check your measurements to make sure they are correct. Cut off any extra tubing.

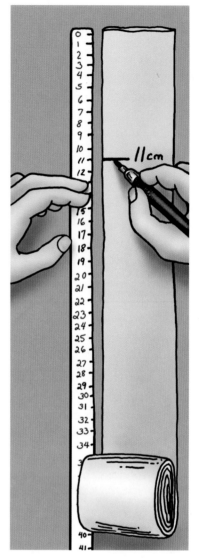

Figure 2.2 *How to mark the tubing*

Table 2.1 Length of the Organs of the Digestive Tract

Organ	Length (cm)
Mouth	11
Esophagus	25
Stomach	22
Small intestine	690
Large intestine	152
Rectum	14

4. Using the marker, label each space on the tube with the name of the appropriate organ.

5. Pick up the tube, the plastic bag with the tennis ball, and the small empty plastic bag. Move with your group to an uncrowded area of the room. Have all group members space themselves out at equal distances along the length of the tube. The first person in the group should have the bag with the ball, and the last person in the group should have the empty bag.

6. Have the student who is holding the end of the tube that is marked "Mouth" squeeze the tennis ball from the plastic bag into the end of the tubing, as illustrated in Figure 2.3. Squeeze the ball gently into the tube. Be careful not to touch the greasy ball!

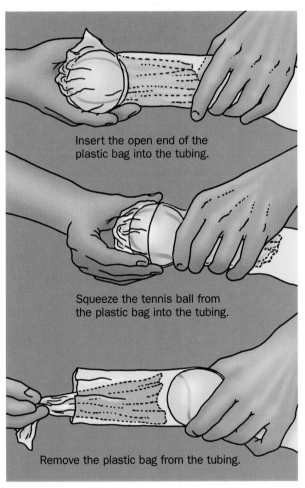

Insert the open end of the plastic bag into the tubing.

Squeeze the tennis ball from the plastic bag into the tubing.

Remove the plastic bag from the tubing.

Figure 2.3 *How to put the ball into the tube*

7. Taking turns, move the tennis ball through the tube. Be sure to keep the tube horizontal at all times. Give everyone a chance to squeeze. While you are squeezing, think about the following questions:

A. What does the tennis ball represent?

B. Why do you think the tennis ball was soaked in oil?

8. Continue squeezing until the ball pops out of the tube into the plastic bag.

9. Follow your teacher's instructions for cleanup.

REFLECTING ON WHAT YOU'VE DONE

1. With your group, revisit the questions you answered at the beginning of this inquiry and the questions in Step 7 of the Procedure. Make revisions if necessary. Share your thoughts with the class.

2. How was the movement of the tennis ball similar to the movement of food as it passes through the digestive tract? How was it different? Discuss these questions with the class.

3. What would you like to know about your digestive system? About nutrition or diet? Make a list in your science notebook. You'll revisit these questions at the end of this part of the module.

ALL SYSTEMS GO!

Introduction

In this module, you're going to explore something that's very familiar to you: the human body. You will be learning about the digestive system, the respiratory system, the circulatory system, and the musculoskeletal system.

Even though you already know a good deal about your body, you are probably in for a few surprises. Learning about your body and how it functions is important, because it can give you the information you need to take good care of your health.

In class, you'll be doing investigations to help you understand how the body functions. For

Peppi and Bollo take a look at their destination: planet Earth.

example, you've already begun to explore how food moves through the digestive tract.

Science Fiction That's Loaded With Facts

You will also be reading about the organs and systems of the body. Some of the reading selections in this module are part of a series entitled "Spies." In this series, you'll follow the adventures of Peppi and Bollo, two creatures who have been sent to planet Earth by scientists in another galaxy. Peppi is in charge. Bollo is her pupil at the Intergalactic University of Saganova. His main job is to ask questions and take good notes.

Once they've completed their mission, Peppi and Bollo will return to mission control. The scientists who hired them want to learn everything they can about the human body. Then, the scientists hope to use all their data to design a perfect human body.

Peppi and Bollo have the ability to make themselves (and anything they are wearing or carrying) very small—so small that they can travel through a human body without being detected! They're well equipped for the journey. In their backpacks are powerful magnifying lenses and other scientific equipment. They are also carrying a famous textbook called *Gray's Anatomy* (just in case Peppi has a lapse of memory).

Peppi and Bollo have a little more work to do to prepare for their journey. You'll see them again in Lesson 4. Until then, enjoy these introductory lessons that will prepare you for your own journey through the human body. ☐

Bollo won't have time to play soccer or listen to CDs on his mission. But he will need some scientific equipment (and maybe a toothbrush?).

Exploring Carbohydrates

© YOAV LEVY/PHOTOTAKE/PNI

Nutrition Facts
Serving Size 3/4 cup (140g prepared)
Servings Per Container about 4

Amount Per Serving	Dry Mix w/ water	Dry Mix w/ 2 Tbsp butter
Calories	160	210
Calories from Fat	5	60

	% Daily Value**	
Total Fat 0.5g*	1%	10%
Saturated Fat 0g	0%	19%
Cholesterol 0mg	0%	5%
Sodium 750mg	31%	34%
Total Carbohydrate 34g	11%	11%
Dietary Fiber 1g	4%	4%
Sugars 1g		
Protein 4g		

Vitamin A	15%	15%
Vitamin C	6%	6%
Calcium	4%	4%
Iron	2%	2%

*Amount in Dry Mix prepared with butter. One tablespoon butter contributes an additional 25 calories, 30mg sodium, 0g total carbohydrate, (0g sugars), and 0g protein per serving.
**Percent Daily Values (DV) are based on a 2,000 calorie diet. Your daily values may be higher or lower depending on your calorie needs:

	Calories:	2,000	2,500
Total Fat	Less than	65g	80g
Sat Fat	Less than	20g	25g
Cholesterol	Less than	300mg	300mg
Sodium	Less than	2,400mg	2,400mg
Total Carbohydrate		300g	375g
Dietary Fiber		25g	30g

Calories per gram:
Fat 9 • Carbohydrate 4 • Protein 4

INGREDIENTS: PARBOILED LONG GRAIN RICE, SALT, DEHYDRATED RED PEPPERS, ONION POWDER, TOMATO POWDER, YEAST EXTRACT, PAPRIKA, GARLIC POWDER, SPICES, EXTRACTIVE OF ANNATTO, MODIFIED CORN STARCH DOES NOT CONTAIN ANIMAL PRODUCTS

Nutrition Facts labels contain important information. Read before you eat!

INTRODUCTION
In Lesson 2, you explored how food moves through your digestive tract. But how is food digested? How does your digestive system change that sandwich and apple you ate for lunch into a form that your body can use to grow and stay healthy? This is what you will be investigating in the next few lessons.

Before you can begin these investigations, you need to be able to perform chemical tests for sugar and starch. Sugar and starch are carbohydrates, one of the three major types of food. The tests you do will help you decide which of these nutrients are in the foods you eat. Then you will be ready to see whether these nutrients change as a result of the digestive processes that take place in the mouth.

OBJECTIVES FOR THIS LESSON

Indicate what carbohydrates are and the forms in which they are found.

Perform chemical tests to determine the presence of sugar and starch in five foods.

Explain why the human body needs carbohydrates.

WHAT IS A WATER BATH?

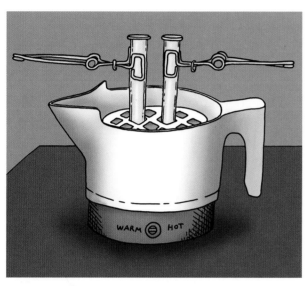

Test tubes resting in a water bath

A water bath is a convenient way to heat a substance when you want to maintain a constant temperature. A water bath is often used in science class because of its safety. If you put a test tube directly into a flame, the liquid in the tube might begin to boil and bubble out of the test tube. This would probably interfere with the test results. It could also cause a serious burn.

Although a water bath is a safer method of heating than a direct flame, it is still very hot. Be careful when working with a water bath.

MATERIALS FOR LESSON 3

For you

1 copy of Student Sheet 3.1a: Food-Testing Data Sheet
1 copy of Student Sheet 3.1b: Venn Diagram: Carbohydrates
1 pair of safety goggles

For your group

1 tote tray
 Samples of the following foods:
  Sugar solution, in dispensing bottles
  Small marshmallows, in package
  Potato buds
  Powdered egg white
  Starch
 Distilled water (DW), in large dropper bottle

2 dropper bottles of Benedict's solution
2 dropper bottles of Lugol solution
2 medium test tubes
2 test tube clamps
1 test tube rack
2 6-cup lab trays
3 scoops
8 toothpicks
2 black markers
2 plastic pipettes

SAFETY TIPS

Wear safety goggles whenever you are working with chemicals or hot plates.

Treat the foods you will use in this lesson as laboratory chemicals. Do not eat them or put them in your mouth.

Do not hold the attached test tube clamp while the test tube is in the hot water bath. Allow the test tube to rest in the boiling water with the clamp attached.

To remove the tube from the hot water bath, grasp the clamp and lift out the tube. Do not touch the test tube. Keep the clamp attached. Rinse the test tube with cool water. When the test tube is cool, remove the clamp. Clean the tube with the brush.

If Benedict's solution comes into contact with your eyes, rinse them thoroughly with water. If the solution touches your skin, wash it with soap and water.

Lugol solution is poisonous. Do not allow it to touch your mouth, nose, skin, or eyes. Notify your teacher immediately if this occurs. Lugol solution will also stain your skin and clothing.

Report any spills or other accidents with the chemicals to your teacher immediately. Your teacher will explain how to clean up the spills.

Getting Started

1. Listen as your teacher talks about how to perform chemical tests for sugar and starch.

2. You will use two chemicals to perform these tests. The chemicals are harmless when they are handled properly. All chemicals, however, must be used with great care. Review the Safety Tips with your teacher.

3. Divide your group into pairs. You and your partner will test five foods for sugar and for starch. You will also test distilled water. Perform the chemical tests for sugar first. Then, using the same food samples, perform the chemical test for starch.

Inquiry 3.1
Testing Foods for Sugar and Starch

PROCEDURE

1. Pick up your materials.

2. Using Figure 3.1 as a guide, number your lab cups from 1 to 5. Label the remaining cup "DW" (for Distilled Water). Number the scoops "3," "4," and "5." You will use a different scoop for each of the solid foods.

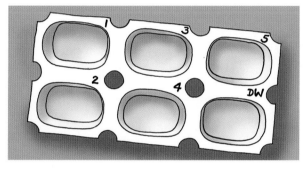

Figure 3.1 *How to label your lab tray*

3. Add substances to the cups in the following manner:

A. Cup 1: Add 30 drops of sugar solution.

B. Cup 2: Add one marshmallow and 40 drops of distilled water.

C. Cups 3, 4, and 5: Use the scoops numbered 3, 4, and 5 to put a small amount of foods 3, 4, and 5 in the similarly numbered cups. Use enough food to cover the bottom of the cup. Be sure that each scoop has the same number as both the cup and the solid food. Add 40 drops of distilled water to each cup.

D. Cup DW: Add 30 drops of distilled water.

E. Stir cups 2, 3, 4, and 5 with different toothpicks.

4. Test the six substances for sugar as follows:

A. Fill the pipette with distilled water from the cup labeled DW. Add 20 drops of the distilled water to a test tube. Return the remaining distilled water to the cup.

B. Add 10 drops of Benedict's solution to the test tube.

C. Attach the clamp to the test tube. Grasp the clamp. Place the test tube, with clamp attached, into the water bath (see Figure 3.2 for one type of water bath). Let it sit for 1 minute.

Figure 3.2 *These students are performing the inquiry with a different type of water bath.*

D. Using the clamp, remove the test tube from the water bath. Observe the color of the solution and record it in the appropriate blank on Student Sheet 3.1a. Refer to Figure 3.3 to decide whether the sample contains sugar. If sugar is present, put a "+" sign in the next column. If sugar is not present, put a "−" sign in that space.

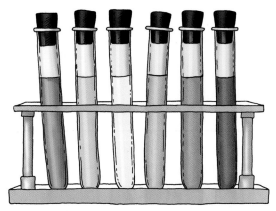

Figure 3.3 *The colors of the substances in the test tubes range from light green to red, depending on the concentration of sugar. The concentration of sugar is lowest in the light-green substance and highest in the red substance. Use the colors of the solutions in these test tubes to interpret your test results.*

E. Rinse the test tube and pipette well with cold water. Remove the clamp. Clean the tube thoroughly with the brush, and soap and water.

SAFETY TIP

When the test tube is cool, remove the clamp. Then clean the test tube. Otherwise, the test tube might slip out of the clamp and break while you are cleaning it.

5. Repeat these testing and cleaning procedures for each of the five foods in your tray. Record your results on your student sheet.

6. When you have finished your tests for sugar, have one member of your group record your results in the appropriate column on the Food-Testing Master Chart. If there is no color change, write "blue" in the space. Place a "+" sign in the next column if sugar is present; place a "−" sign if sugar is not present.

7. Now test the foods for starch according to the following procedure.

A. Add 3 drops of Lugol solution to each cup in your lab tray (as shown in Figure 3.4).

Figure 3.4 *You will use Lugol solution to test for starch.*

B. Observe the color in cups 1 through 5. If the food does not dissolve in the distilled water, look for a color change in the food itself, not in the liquid.

C. For an example of a negative test, check the color of the Lugol solution and distilled water in cup 6.

D. Record your findings on the student sheet.

E. When everyone in your group has finished the tests, share your results.

F. Have one group member record the color of each of the foods and a "+" or "−" sign on the Food-Testing Master Chart.

8. Follow your teacher's directions for cleanup.

REFLECTING ON WHAT YOU'VE DONE

1. Discuss the results on the Master Chart with your class.

2. Share your results from Student Sheet 3.1b: Venn Diagram: Carbohydrates with your class.

NUTRIENTS:
You Just Can't Live Without 'Em

An Indian family enjoys dinner.

"Do you want fries with that cheese-burger?"

"Sure, and a large soda, please."

You've ordered a great-tasting lunch, but is it a healthy meal? The good news is that you're getting many of the nutrients you need. The bad news is that you're also about to eat a lot of stuff that you don't need—like extra salt and fat.

When you choose what to eat, you need to think about more than your taste buds. You need to know what your body needs and what foods will give it to you. You need to know about nutrients.

What are nutrients? They are the fuels your body needs to keep you going. They are used for growth and repair. They help fight disease.

There are six types of nutrients: carbohydrates, proteins, fats, vitamins, minerals, and water. The first three—carbohydrates, proteins, and fats—make up the three basic food types.

Foods are part of cultural traditions. All foods, however, contain one or more of the same essential nutrients. Can you tell which of the foods that you see in the first three photographs of this article are high in carbohydrates? Fat? Protein?

Tortillas, beans, and trimmings are popular foods in Mexico and many other countries.

Rice and vegetables are mainstays of this Asian family's diet.

Carbohydrates: Energy to Burn

The two most important kinds of carbohydrates are sugars and starches. The simplest carbohydrate is a sugar called glucose. Glucose is your body's first choice for fuel.

Complex carbohydrates are called starches. When you eat complex carbohydrates such as pasta, bread, or rice, your body breaks them apart to get the simple sugars that it needs for fast fuel.

But your body doesn't waste the leftovers. It turns most of the extra glucose into a substance called glycogen and saves it in your muscles and liver. If you eat more food than you need at the moment, some of the glucose gets turned into fat. Your body stores fat to make sure it has fuel for its future energy needs. (It's like putting the fat away in a warehouse.)

But watch out: the warehouse can get too full. If you continually eat too much starch or simple sugar—such as the sugar in corn syrup, candy, and soft drinks—you can quickly put on extra pounds. What's more, candy bars and soft drinks usually don't contain other nutrients, like minerals and vitamins that you can get from fruits and vegetables.

Fiber is another carbohydrate. It is made of the same raw materials as sugars: carbon, hydrogen, and oxygen. But you can't digest fiber. So what good is it? Fiber provides bulk that helps move food through your intestines. Fruits provide fiber called pectin. Whole-grain breads and cereals contain cellulose, the fiber that forms a plant's cell walls.

Proteins: The Body Builders

Proteins are at work in every cell of your body. You need proteins to build tissue (muscle tissue contains a lot of protein). Your body also uses proteins to repair damage and to make substances such as hemoglobin, which carries oxygen through your blood.

When your body digests food, it breaks the protein down into simpler substances called amino acids.

These foods are good sources of carbohydrates. Can you name them? Do you know any other foods that are high in carbohydrates?

For many people, meat is an important source of protein. Other protein-rich foods include nuts, egg whites, and cheese.

You might compare amino acids with letters in the alphabet. There are 20 amino acids, and they combine to create millions of different protein "words." Your body also builds some amino acids from scratch by combining carbon, hydrogen, oxygen, and nitrogen.

There are nine amino acids that you can get only from the food you eat. Your body can't build them on its own. These are called the essential amino acids. A food that has all nine essential amino acids is called a complete protein. Food that comes from animals (for example, meat, eggs, and milk) has complete proteins.

Foods that do not have all the essential amino acids are called incomplete proteins. Foods that come from plants—for example, vegetables, fruit, wheat, and rice—contain incomplete proteins.

If you don't want to eat meat or animal products, you can still get your essential amino acids. Just make sure you combine foods that have *different* incomplete proteins. For example, a meal that combines black beans and rice, or lentil soup and corn bread, will give you the complete proteins that you need.

Fats: What's the Skinny?

Too much fat can cause health problems such as obesity and heart disease. But fat isn't all bad news. It has more energy than proteins or carbohydrates. A gram of fat supplies about twice as much energy as a gram of protein or carbohydrates.

What else is good about fat? Thin layers of fat act as protective padding around your heart and other organs. Fat helps insulate your body, too, so you don't have to burn too much fuel to stay warm.

There are two kinds of fats: saturated and unsaturated. Most fats that come from animals are saturated fats. Inside your body, saturated fats can turn into cholesterol—a substance that may collect inside your blood vessels and place an extra burden on your heart. Most fats that come from plants (for example, from nuts and vegetables) are unsaturated fats.

Even though your body can convert carbohydrates into fat, you still need to eat some foods that contain fat. Why? Because your body can't build some of the unsaturated fats that you need.

But be careful. You need only a small amount, and the fatty foods you eat should contain mostly unsaturated, rather than

Eat only small quantities of foods such as these, which are high in fat.

saturated, fat. Meat, cheese, and butter have saturated fat. Sweet baked goods like cookies and cakes also have saturated fat. Fish, avocados, and most liquid cooking oils are sources of unsaturated fats.

Vitamins: The ABCs of Health

Vitamins are chemicals that have been made by living organisms. Scientists discovered the first vitamin (B-1, or thiamin) about 100 years ago. They have now identified a total of 13 vitamins. Each of these vitamins has an essential role in the chemical reactions that go on in our bodies. Vitamins help build blood cells and chemicals that control the nervous system.

You only need tiny amounts of vitamins, but your body can't manufacture them. The best sources of most vitamins are fresh fruits and vegetables.

Minerals: Little Things That Mean a Lot

Minerals are chemicals that occur naturally in the environment. They do not need to be made by a living organism. You need minerals to build bones, teeth, and blood cells. Minerals also regulate the chemical and electrical signals that control the way your body works. Minerals come from the earth. They are absorbed by plant roots as they grow, and they are passed on to animals that eat plants.

You get minerals from many foods—fruits, vegetables, and grains, as well as meat and milk.

You need only tiny amounts of some minerals (for example, copper, iodine, iron, and zinc). These are called trace minerals. You need more of other minerals such as calcium, magnesium, potassium, and sodium. These are the macrominerals.

Calcium, found in milk and cheese, is especially important for children who are still growing. Your body uses calcium to build the hard, strong parts of your body—bones and teeth, for example. Older people need calcium, too, to keep bones strong. Iron is a mineral component of hemoglobin. Because women lose blood during their menstrual cycle, they need more iron than men. Beef, tuna, and chicken are good sources of iron.

Water: Where It All Comes Together

How can water be a nutrient? There's nothing in it! There may not be much nutrition in water, but there's an awful lot of water in you. Your brain and muscles are three-fourths water, and bone is 20 percent water.

Every cell in your body is packed with water. Your body needs it to transport nutrients and wastes, control your temperature, and carry out chemical reactions.

Each day, your body loses more than 2 liters of water. You replace some of it with food, but be sure to drink six to eight glasses of liquid each day to maintain your water supply. When you're thirsty—and even when you're not—have some water.

So What's for Dinner?

The best way to get the nutrients you need is to eat good, fresh food. Get enough of the right stuff, and you won't have to worry about vitamin or mineral pills. Knowing about nutrients can help you choose foods that give your taste buds what they want and the rest of your body what it needs. □

Copomer is a trace mineral. You need just a tiny bit of it in your diet. In fact, if this penny were pure copper, it would contain enough of this mineral to meet your daily needs for three and a half years!

Digestion in the Mouth

© WILL/DENI McINTYRE/PHOTO RESEARCHERS, INC.

That pizza looks great! But take it easy—it's important to chew your food carefully before you swallow it.

INTRODUCTION

Now that you know how to perform chemical tests for starch and sugar, it's time to use those skills to begin your investigation of the digestive processes. In this lesson, you will see what happens to food while it is in the mouth.

Is there a good reason to chew your food well before you swallow it? Do your teeth get any help as they tear and grind a bite of food before you swallow it?

Once you've explored these questions, you'll be prepared to move on to investigate what happens in the stomach, where the process of digestion continues. All these activities have one goal: to get food into a form that can be absorbed into your bloodstream and used by the cells to maintain life.

OBJECTIVES FOR THIS LESSON

Explore mechanical digestion in the mouth.

Explore chemical digestion in the mouth.

Begin to construct a definition of the word "enzyme."

Getting Started

1. Your teacher will give you a cracker. Put it in your mouth. Chew it very slowly. Pay close attention to everything that happens in your mouth. When you are ready to swallow it, put your thumb and index finger gently around your Adam's apple. Swallow the cracker. In your science notebook, describe exactly what happened while you were chewing and swallowing the cracker.

    A. Why do you think that chewing the cracker was important?

    B. What happened to your Adam's apple as you swallowed?

2. Share your ideas with the class.

Inquiry 4.1
Exploring Chemical Digestion in the Mouth

PROCEDURE

1. Pick up your materials. You will work in pairs in this lesson. You will perform chemical tests for sugar and starch using Benedict's solution and Lugol solution. Take turns so that each of you gets equal experience with both tests.

For you
  1 copy of Student Sheet 4.1: Chemical Testing for Sugar and Starch
  1 sheet of four summary boxes (for homework assignments)
  1 pair of safety goggles

For your group
  1 tote tray
  2 dropper bottles of Benedict's solution
  2 dropper bottles of Lugol solution
  2 dropper bottles of salivary amylase
  2 dropper bottles of distilled water (DW)
  2 dispensing bottles of starch solution
  2 medium test tubes
  1 test tube rack
  2 test tube clamps
  2 6-cup lab trays
  2 plastic pipettes
  8 toothpicks
  2 black markers

SAFETY TIPS

Always wear safety goggles when using chemicals and hot pots.

Lugol solution contains a small amount of iodine, which is considered poisonous. Do not place any of the chemicals in or near your mouth.

Do not get the chemicals on your skin or clothing. Lugol solution will leave a stain. If Benedict's solution comes in contact with your eyes or skin, thoroughly flush the affected area with a generous amount of water.

Wipe up accidental spills of Lugol and Benedict's solution with paper towels, which you can dispose of in the trash.

2. Read along as your teacher reviews the Safety Tips.

3. Number the cups in your lab tray from 1 to 6. Use Figure 4.1 as a guide.

4. Using Figure 4.1 and your teacher's guidance, complete the design of the data collection table on Student Sheet 4.1. Give the table a name.

Figure 4.1 *Designing the data collection table*

5. Add 30 drops of starch solution to cups 1, 2, 5, and 6.

6. Add 30 drops of distilled water to cups 3 and 4.

7. Add 2 drops of amylase to cups 3, 4, 5, and 6, as shown in Figure 4.2.

Figure 4.2 *Add amylase to cups 3 through 6.*

8. Stir the contents of cups 3, 4, 5, and 6 with separate toothpicks. Let the cups rest for at least 5 minutes. While you are waiting, think about the following questions and discuss them with your lab partner.

A. **What do you think amylase is?**

B. **What do you think amylase will do when you mix it with the starch solution?**

C. **Why do you need to wait 5 minutes before you perform the tests on the starch solution and amylase mixture?**

9. Using the technique you learned in Lesson 3, perform a chemical test for sugar for the substances in the cups in the top row (cups 1, 3, and 5). If you don't remember how to perform this test, reread the Procedure for Inquiry 3.1 on pages 17–19.

10. Using the technique you learned in Lesson 3, perform a chemical test for starch for the substances in the cups on the bottom row (cups 2, 4, and 6). Refer to the Procedure in Inquiry 3.1 on pages 17–19 if necessary.

11. Record the results of your tests in your data table on Student Sheet 4.1.

12. Follow your teacher's directions for cleanup.

REFLECTING ON WHAT YOU'VE DONE

1. Discuss your answers to the questions that appear in Step 8 of the Procedure.

2. Using the observations you have made in this inquiry, along with the information in "Spies: Into the System," which appears at the end of this lesson in the Student Guide, finish answering the questions on Student Sheet 4.1.

3. Do you think that anything other than eating might cause saliva to form? Your teacher will tell you about a discovery made by a Russian scientist named Ivan Pavlov. Discuss Pavlov's ideas with the class.

4. Think about what happened when you swallowed the cracker at the beginning of the lesson. What route did it take to get to your stomach? Study Figure 4.3 while your teacher discusses what happens during swallowing.

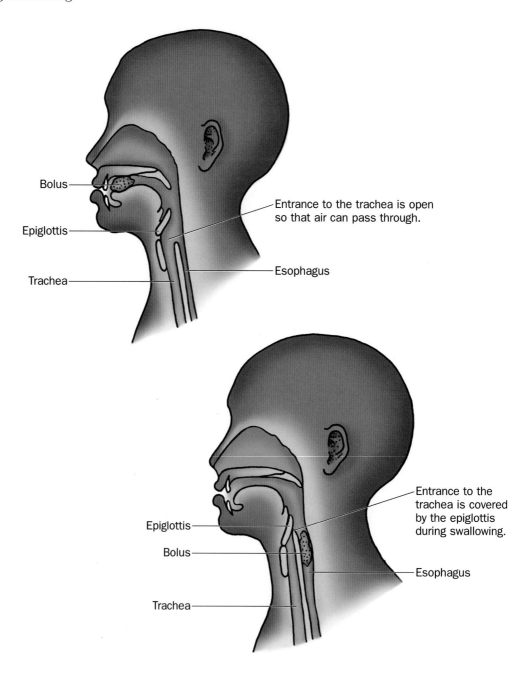

Bolus

Epiglottis

Trachea

Entrance to the trachea is open so that air can pass through.

Esophagus

Epiglottis

Bolus

Trachea

Entrance to the trachea is covered by the epiglottis during swallowing.

Esophagus

Figure 4.3 *Action of the epiglottis during swallowing*

Making It Simple

Why does your body need to break down starch into sugar?

The reason is that the body cannot use carbohydrates until they are in a form that is simple enough to dissolve in water and pass through cell membranes into the blood and, ultimately, to the body's cells.

Starches are made up of chains of simple sugar particles that are held together by chemical bonds.

But once the starch is chewed and acted on by digestive enzymes in the mouth and small intestine, the bonds that hold the links of the starch chain together break apart. The result is simple sugar particles.

These particles are small enough to move through the body and enter the cells. They provide cells with nutrients they need to do their jobs.

Simple sugars. A simple story. Do you agree? ☐

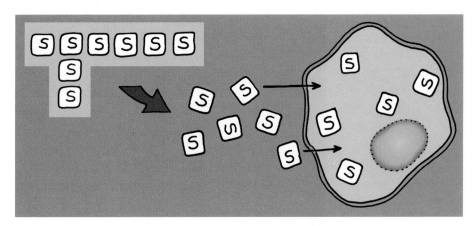

After the chemical bonds in the starch solution are broken, the sugar particles can pass through the cell membrane.

INTO THE SYSTEM

Peppi and Bollo are ready for their first investigation. It's noon. Time for lunch. They hop aboard a slice of pizza (topped with mushrooms, sausage, and green pepper) and enter a human mouth.

"Who's our host?" says Bollo.

"Her name is Joanne. She's 18—an adult as far as human digestive systems are concerned. She's healthy, too. No digestive problems. If we can discover what's going on in Joanne's digestive tract, we'll have a good idea about human digestive activity," replies Peppi.

"Wow, it's dark," says Bollo. "And wet and slippery, too. Watch out for those huge white blades!"

"The liquid is spit," says Peppi. "Its formal name is 'saliva.' Saliva is produced by three pairs of glands in the sides and the back of the human mouth. The salivary glands go to work whenever a human smells, sees, or even thinks about food. Once food is in the mouth, the saliva moistens it and makes it easier to handle. The salivary glands produce about 1.5 liters of saliva each day.

"And those 'blades,'" Peppi goes on, "are teeth. Adult humans usually have 32 of them."

Peppi and Bollo take a look around. It's a busy scene. The front teeth, or incisors, go to work first. These sharp teeth act like scissors. The canine teeth, at the sides of the mouth, are pointy. They cut up the food some more.

The tongue is muscular. It helps move food around. It moves the pizza to the back of the mouth, where the heavy-duty work

Watch out, Joanne! You are about to be invaded by aliens!

of chewing goes on. The broad, flat bicuspids and molars grind the food and make it soft. Meanwhile, the saliva is doing its thing. It moistens the food and makes it easier to chew.

"This tearing, grinding, and mixing," Peppi says, "is called mechanical digestion. And although we can't see it, it's important to know that saliva contains an enzyme called amylase. The enzyme activity marks the beginning of chemical digestion."

"Hold on a minute. You're going too fast," says Bollo. "What's an enzyme?"

"An enzyme is a special protein produced by the body. Digestive enzymes help the body break down nutrients into forms that the body can use. Amylase, for example, helps break down the starch in pizza to simple sugar. As we continue, we'll see other enzymes that help digest other types of nutrients."

"The human mouth is efficient," notes Bollo. "It's only been a few seconds, and I can't even recognize

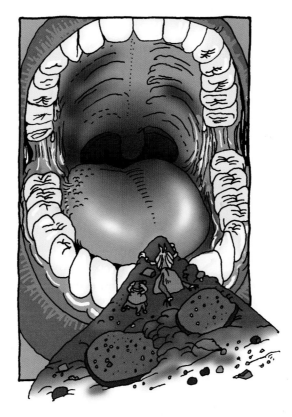

Peppi and Bollo are off to a scary start.

that bite of pizza. It looks like a big, soggy ball."

"You're right. That ball of food is called a bolus," says Peppi. "And if I took out my testing equipment, I could show you that the pizza crust is already starting to be digested—because of the action of amylase."

Down the Tubes

"Eeeeeeee!" shrieks Bollo. Before he can say more, he and Peppi are pushed up against the roof of the mouth. It's a tight squeeze.

Then they start moving backward. The bolus has triggered a swallowing reflex. Goodbye, mouth!

Looking down, they see something close over one of the two tubes below them.

"The epiglottis," says Peppi. "Each time a human swallows, it closes over the windpipe to make sure that food goes to the stomach. The windpipe leads to the lungs, and you don't want any food in there!"

"One tube for air, one for food. The

human body is specialized," says Bollo.

"You're right," says Peppi. "You're also going to find out that these specialized organs and systems need each other. They work together to help keep the body in balance. Now keep your eyes open. We're traveling through the second organ of the digestive system, the esophagus."

Down and down they go, squeezed by muscular contractions of the walls of this dark-pink tube. Then all of a sudden, another tight squeeze, and pop!

The spies arrive in the stomach. They have passed through the last gatepost on the road to the stomach, the lower esophageal sphincter. The sphincter is a ring of muscle that helps keep food that has been swallowed where it should be—in the stomach.

"What happens now?" asks Bollo.

"I like your curiosity, Bollo. Stay posted. Our journey will soon continue," Peppi replies. □

5
Digestion in the Stomach

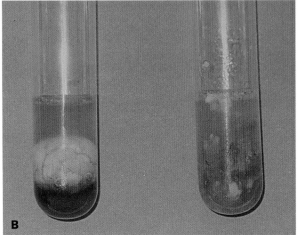

Both of these photographs show the same two test tubes. Each test tube contains a liquid and a small piece of ground beef. Photo B was taken 24 hours after Photo A. What do you think has happened to the contents of each test tube over the 24-hour period?

INTRODUCTION

In Lesson 4, you discovered that the teeth and tongue have major roles in mechanical digestion. You also found out that starch is the only type of food that starts to undergo chemical digestion in the mouth.

In the stomach, as you will discover in this lesson, there is even more digestive action than in the mouth. Three layers of stomach muscles churn the food and continue the process of mechanical digestion. Chemical digestion becomes even more important as the stomach continues to break down foods into a form that the body can use.

What special features help the stomach break down protein, one of the three basic food types? You're about to find out!

OBJECTIVES FOR THIS LESSON

Explore chemical digestion in the stomach.

Discover the roles of hydrochloric acid (HCl) and pepsin in chemical digestion in the stomach.

Explain why the human body needs protein.

Getting Started

1. In your science notebook, describe what you know about what happens to food in the stomach.

2. Share your ideas with the class.

Inquiry 5.1

Exploring Chemical Digestion in the Stomach

PROCEDURE FOR PERIOD 1

1. Your teacher will go over the Procedure for this inquiry. Think about, and be prepared to answer, the following questions:

   A. What will be the control for this lesson?

   B. How do you calculate the volume of a solid object?

   C. Which of the major food types is egg white?

MATERIALS FOR LESSON 5

For you
   1 copy of Student Sheet 5.1: Chemical Digestion in the Stomach
   1 pair of safety goggles
   1 blank summary box (for the human body systems poster)

For your group
   1 tote tray
   1 dropper bottle of HCl
   1 dropper bottle of pepsin
   1 dropper bottle of gastric juice
   1 black marker
   3 medium test tubes
   3 pieces of string
   3 pieces of egg white
   1 30-centimeter (cm) ruler
   1 10-milliliter (mL) graduated cylinder
   1 6-hole test tube rack

SAFETY TIPS

Always wear safety goggles when you are using chemicals.

Treat the egg whites that you will use in this lesson as you would treat lab chemicals. Do not put them in your mouth.

Always handle chemicals with great care. The three chemicals you will use in this inquiry are safe when handled correctly. However, if you get them in your mouth or if they come into contact with your eyes or skin, they can be dangerous. If any of the chemicals touches your eyes or skin, rinse it off thoroughly with water. HCl and artificial gastric juice may also damage clothing.

Report any spills or accidents to your teacher immediately.

2. Review the Safety Tips.

3. Complete the data table that has been begun for you on Student Sheet 5.1.

4. Pick up your materials and begin your inquiry.

5. Use a marker to label your test tubes 2, 3, and 4. Print your initials or your group's name on them. Place the test tubes in numerical order in the test tube rack.

A. Measure 4 mL of pepsin with your graduated cylinder and pour it into Test Tube 2 (see Figure 5.1).

B. Measure 4 mL of HCl with your graduated cylinder and pour it into Test Tube 3.

C. Measure 4 mL of gastric juice with your graduated cylinder and pour it into Test Tube 4.

D. Double-check the three test tubes to make sure the levels of liquid are accurate.

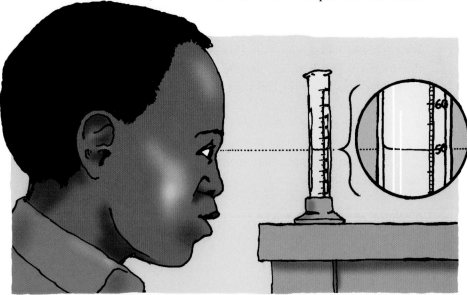

Figure 5.1 *How to read the level of liquid in the tube. Place the graduated cylinder on a level surface, move your head until the meniscus is at your eye level, and read the scale at the bottom of the meniscus.*

6. Use a ruler to measure the length, width, and height of a piece of egg white. Record all the dimensions in your data table. Use these dimensions to calculate the approximate volume of each piece. Record the volumes in your data table.

7. Make a loop with the end of a piece of string, as shown in Figure 5.2. Tighten the string around the egg white until you can lift it by the string. Do not pull too tightly.

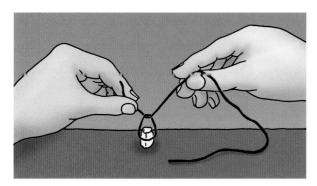

Figure 5.2 *How to tie the string around the egg white*

8. Holding the end of the string, lower the egg white into Test Tube 2 until it is immersed in the pepsin. Let the remaining length of string rest along the outside of the test tube.

9. Repeat the same steps for the other two pieces of egg white. Place those pieces in Test Tubes 3 and 4.

10. When you have set up your test tubes, check to make sure they are labeled with your group's name. Then give them to your teacher, who will place them in a water bath at 37 °C for at least 45 minutes. Now think about the following questions:

A. Why is it necessary to heat the test tubes to 37 °C?

B. Why do you think it was not necessary to heat the crackers and amylase to 37 °C in Lesson 4?

11. Add any details that you think are necessary to your data table on Student Sheet 5.1.

12. Follow your teacher's directions for cleanup.

PROCEDURE FOR PERIOD 2

1. With your class, discuss your answers to the two questions in Step 10 of the Procedure for Period 1.

2. When your teacher returns your test tubes, follow your teacher's directions for completing the inquiry. Make the necessary measurements and record your data.

3. Clean up your work area and return your materials.

REFLECTING ON WHAT YOU'VE DONE

1. On the basis of what you've learned from Inquiry 5.1, class discussion, and your reading, complete the answers to the questions on Student Sheet 5.1 and then discuss your results with the class. Under what conditions did the egg white break down most? The least? What chemicals must be present to break down proteins in the stomach? How does mechanical digestion occur in the stomach? What is the function of the mucus that coats the stomach wall? Why is the term "heart-burn" inaccurate? What is a complete protein? What are some foods that contain complete proteins? How do our bodies use proteins?

2. Take another look at the two photos on the first page of this lesson. Discuss the following questions with the class.

   A. What nutrients are in ground beef?

   B. How did the appearance of the contents of each test tube change over 24 hours?

   C. In which test tube do you think digestion took place? Why?

   D. Which nutrient showed evidence that it cannot be digested by gastric juice?

3. Suppose that you eat lunch at noon every day. Today, you've been extremely busy. It's 3:00 P.M., and you still haven't had time to eat. What might be happening in your stomach? Why? Discuss this with the class.

Chance of a Lifetime

Dr. William Beaumont

William Beaumont was a young frontier doctor living in northern Michigan in the early 1800s. Most of his work was fairly routine. Because he lived in a wilderness area where hunting was common, he had even gotten accustomed to treating gunshot wounds.

But the wound that Alexis St. Martin, a young fur trapper, had suffered was different. The shot blasted a deep hole in his stomach. Beaumont was convinced that St. Martin would not live through the night.

Much to Dr. Beaumont's surprise, his patient did live. The deep wound in his stomach, however, never healed over.

To keep food from leaking out of the wound, Beaumont covered it with a bandage.

Beaumont had always been interested in research, and he realized that St. Martin gave him a rare opportunity to study how the stomach works. So Beaumont persuaded St. Martin to stay with him and let him perform some experiments. Here's what he did.

He took samples of partially digested food from St. Martin's stomach and examined them. He tied threads around pieces of fresh food and dropped them into the stomach. After a while, he pulled out the pieces and observed the effects of the gastric juices. He measured the temperature of the stomach when it was empty and full. He even sent samples of St. Martin's gastric juices to Europe to be analyzed by a chemist.

Beaumont published his findings in a book called *Experiments and Observations.* Doctors and scientists around the world who read the book gained a new understanding of the stomach and how it digests food. ☐

Dr. Beaumont's desk and lab equipment might have looked something like this. What chemicals do you think are in the small bottles?

QUESTIONS

1. What do you think Dr. Beaumont found out about digestive activity in the stomach? What food types are digested in the stomach? What types are not?
2. How do you feel about the way Dr. Beaumont used a human subject for his research? Do you think this could happen today?

INTO THE BLENDER

Peppi and Bollo find themselves in a large, hollow area that looks like a J-shaped balloon.

"Just like folds of pink velvet," says Bollo. "What's going to happen here?"

"First of all, we're in for a bath," says Peppi.

"You're right. What's all this liquid sloshing around?" asks Bollo.

"The digestive juices," says Peppi. "One of them is an acid, called hydrochloric acid. The other juice is an enzyme called pepsin. Together, these juices are called gastric juices. They begin the digestion of protein."

"Hydrochloric acid. That's pretty strong stuff, isn't it?" asks Bollo.

"Yes. Pure hydrochloric acid is extremely strong. But the hydrochloric acid in the stomach is weaker. In fact, in the stomach, it's actually the pepsin that does most of the work of digestion. The hydrochloric acid just gets things going."

Three Types of Muscles

"The churning of the stomach helps the digestive juices do their job even better," Peppi continues. "It's caused by muscles. How many can you see?"

"There's one type of muscle that runs at an angle," says Bollo. "But where are the others?"

"Good for you, Bollo. That's one of the three. But there are two more layers of muscle on top of that one. One layer circles the stomach, and another runs from one end of the stomach to the other. All this action helps move the food around and helps break it into tiny pieces."

"It was about 12:30 P.M. when we arrived,"

says Bollo. "That pizza was part of Joanne's lunch. But her stomach wasn't empty. Was that her breakfast?"

"Yes, there are still some small pieces of food in here. Some things take longer for the stomach to process than others. But Joanne's breakfast is already well on its way to becoming digested. It has turned into a thick, creamy liquid called chyme. Chyme is spelled c-h-y-m-e. But remember, it rhymes with 'dime.'"

Peppi and Bollo keep an eye on the action. More food, as well as some liquid, enters. The stomach gradually expands. The thick folds that Peppi and Bollo saw when they entered the stomach are flattening out.

As time passes, the pieces of food get smaller and smaller. The sloshing continues.

Slippery Stuff

"That gastric juice must be strong," says Bollo. "But why doesn't the juice attack the stomach itself? How does it know what's 'off limits'?"

Peppi and Bollo explore the stomach. What do you think is dripping from the top?

"The stomach is protected by an inner wall that is covered with a thick liquid called mucus," replies Peppi. "The stomach generally doesn't start producing digestive juices until food is present and the mucus is in place.

"As long as the stomach is coated with mucus, the gastric juice usually cannot do any harm. But if the juice finds a spot where there is no mucus, it can penetrate the stomach wall. It makes a small hole, called an ulcer."

"So here's a potential problem with this marvelous organ called the stomach," says Bollo, perking up his ears. "It can get ulcers. Something to report to our leader." He jots this down in his notebook.

"Yes. Ulcers have many causes. Scientists recently learned, for example, that they can be caused by certain bacteria as well as by too much gastric juice or too little mucus. But once a doctor has diagnosed an ulcer, it's pretty easy to treat.

"And while we're on the topic, there's another type of discomfort that's caused by gastric juice. The pain develops in the lower esophagus, just around the entrance to the stomach. It's called heartburn. It has nothing to do with the heart; the pain just happens to develop in the same area where the heart is located. Gastric juice backs up into the lower esophagus from the stomach and causes a burning sensation."

"Heartburn," writes Bollo.

"But don't get carried away thinking about the problems that can arise in the stomach. It's a very efficient organ. It even has a way of getting rid of things that don't agree with it. It reverses the normal digestive process, opens that sphincter between the esophagus and stomach, and sends them right back up and out! That action is called vomiting. It is not very pleasant, but absolutely necessary. Vomiting can be triggered by bad food, by some medications, or by poisons."

"I have another question," says Bollo. "Some of the foods seem to be disappearing faster than others. I don't see much pizza crust around others. I don't see much pizza crust around

here anymore, but there's lots of sausage."

"Remember that there are three major food types—carbohydrates, fats, and proteins. Pizza contains all of them. The crust is mostly carbohydrate, and the sausage is fat and protein. All three food types undergo mechanical digestion in the mouth, but only carbohydrates begin to undergo chemical digestion in the mouth."

Why does mucus have to be so slimy? Bollo knows. Do you?

"So that's why I see less crust and more sausage in the stomach!" says Bollo.

"Right," Peppi replies. "Proteins begin to undergo chemical digestion in the stomach. The chemical digestion of carbohydrates continues for a short time in the stomach as well. Fats, on the other hand, are not chemically digested in the stomach at all. It all depends on enzymes. There is even a special enzyme called rennin that begins the digestion of milk. That is important, because when humans are born, that's all they eat for a while."

"But what happens now?" says Bollo. "We've been here for almost 4 hours!"

"Be patient, Bollo. Like the pizza that Joanne ate, we still have a long way to go!" □

LESSON 6

Diffusion and Active Transport

How does the fragrance of flowers reach your nose? You are about to find out!

INTRODUCTION

You've explored the digestive processes that take place in the mouth and stomach. Now it's time to see what happens in the small intestine. Even though a great deal of digestive activity has already taken place, there is a lot more to come. In fact, it's in the first section of small intestine where most chemical digestive activity takes place. As digestive products continue to move through this important digestive organ, most of them are absorbed into the bloodstream.

The activities you will perform in this lesson will help you understand how the process of absorption occurs. First, you will set up an experiment that will show whether sugar and starch can pass through a membrane. While you're waiting to perform the sugar and starch tests, you will do another quick activity that is related to absorption. You will put some anise

OBJECTIVES FOR THIS LESSON

Use models to show how substances tend to spread.

Determine whether certain substances can pass through a membrane by diffusion.

Construct a definition of the word "diffusion."

Explain the difference between diffusion and active transport.

Indicate why both diffusion and active transport are necessary for the absorption of nutrients.

Summarize the digestive processes that take place in the small intestine.

seeds in a balloon and inflate it. Your nose will be your guide to interpreting the results of this test! You'll then put some anise extract in a balloon. Sniff, sniff. Smell the difference!

Getting Started

1. Your nose is going to have an important role in this inquiry! You probably already noticed a strong aroma as soon as you entered the room.

2. Listen as your teacher presents some introductory information. Now discuss the following questions with the class:

A. How do you think the scent reached your nose?

B. Was the aroma stronger at one place than another? If so, why?

3. Do you think that substances can move through solids as easily as they pass through a gas such as air? For example, could you detect an odor through plastic or rubber? Write down your thoughts in your notebook.

4. Before you begin your inquiry, turn in the summary box on which you have recorded the digestive processes that occur in the mouth.

MATERIALS FOR LESSON 6

For you
1 copy of Student Sheet 6.1: Diffusion and Cell Membranes
1 pair of safety goggles

For your group
1 tote tray
1 dropper bottle of Benedict's solution
1 dropper bottle of Lugol solution
1 bottle of sugar solution
1 bottle of starch solution
1 dropper bottle of distilled water
1 plastic pipette
1 250-mL beaker
1 50-mL graduated cylinder
1 plastic funnel
1 test tube clamp
2 large test tubes
1 medium test tube
1 test tube rack
2 membranes
2 pieces of string (4 pieces for the first class)
1 dropper bottle of anise extract
Anise seeds
2 balloons
1 small plastic bag
1 black marker

Inquiry 6.1
Spreading Out and Through

PROCEDURE

1. Have a student from your group get a tote tray. You will work with your group for this inquiry. With the class, go over the Procedure.

2. Before you begin the inquiry, review the Safety Tips.

3. Set up your materials as follows:

A. Label one of the large test tubes "Sugar Solution."

B. Remove a membrane from the beaker. Fold one end of the membrane back about 1.5 cm, as shown here. Make a loop with one end of the string and tie it securely around the folded end of the membrane.

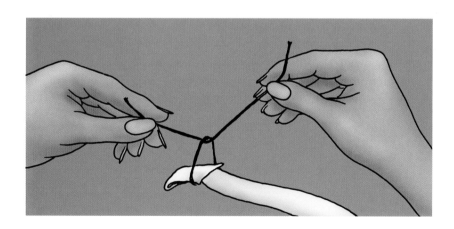

C. Make sure the untied end of the membrane is wet. Then rub it between your thumb and forefinger until it opens, as shown.

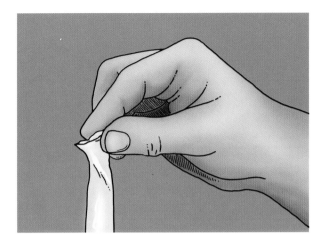

D. Use the graduated cylinder to measure 13 mL of sugar solution. Insert the end of the funnel through the open end of the membrane. Then, as shown in the figure, pour the sugar solution from the graduated cylinder into the membrane.

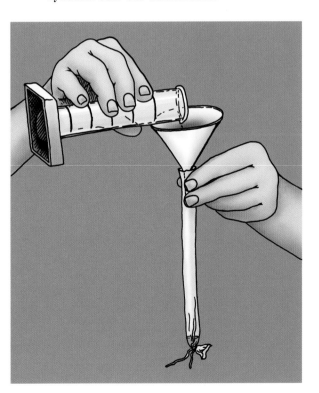

E. Rinse the outside of the membrane thoroughly to remove any trace of the sugar solution.

F. Place the membrane into the test tube labeled "Sugar Solution." Hang the excess membrane over the top and down the side of the tube.

G. Rinse the graduated cylinder. Measure 8 mL of distilled water and pour it into the test tube.

H. Place the test tube in a beaker or test tube rack. The top of the membrane should still hang over the test tube.

4. Label the other large test tube "Starch Solution." Then use steps B through H as a guide in setting up the test tube for the starch solution.

5. Allow both test tubes to sit in a beaker or test tube rack for about 15 minutes, as shown here.

6. While you are waiting, complete this short activity.

A. Remove a balloon from the bag.

B. Open the container of anise seeds. Smell them. What do they smell like?

C. Put about 10 seeds into the balloon. Blow it up. While one person pinches the neck of the balloon to keep air from escaping, have a second person tie the balloon securely with the twine. Pass the balloon around. Let everyone in your group sniff it (see Figure 6.1). Can you smell the seeds through the walls of the balloon? Why or why not?

D. Put 5 drops of anise extract into the other balloon. Inflate the balloon, tie it securely, and pass it around to let everyone sniff it. Can you smell the anise extract through the wall of the balloon? Why or why not?

E. How does the strength of its aroma compare with that of the seeds in the balloon? What might explain the difference? Discuss these questions with your group.

7. Now it is time to turn your attention back to the sugar and starch tests. Perform a chemical test for sugar on the water from the test tube with the membrane containing sugar solution.

A. Use the dropper pipette to add 20 drops of the water in which the membrane is soaking into a test tube.

B. Add 8 drops of Benedict's solution to the test tube.

C. Place the test tube in the hot water bath for 2 minutes.

COURTESY OF HENRY MILNE/NSRC

Figure 6.1 *Do you think that this student can smell the scent of anise seeds through the balloon? What about anise extract?*

8. Perform a chemical test for starch on the water from the test tube with the membrane containing starch solution. Put 10 drops of Lugol solution directly into the water in the large test tube and observe what happens. There is no need to heat the tube.

9. Record your results on your student sheet.

10. Follow your teacher's directions for cleanup.

REFLECTING ON WHAT YOU'VE DONE

1. Use your findings from this inquiry and the information in the two reading selections at the end of this lesson to answer the questions on Student Sheet 6.1. With your class, review Student Sheet 6.1 and clarify the concepts of diffusion and active transport.

2. Suppose someone dropped a small, uncapped bottle of red food coloring into a swimming pool. In terms of diffusion, what would happen?

3. In your science notebook, summarize the steps involved in the completion of digestion in the small intestine. After reviewing Figure 6.2 with your teacher, explain how the accessory organs of digestion aid in the completion of digestion.

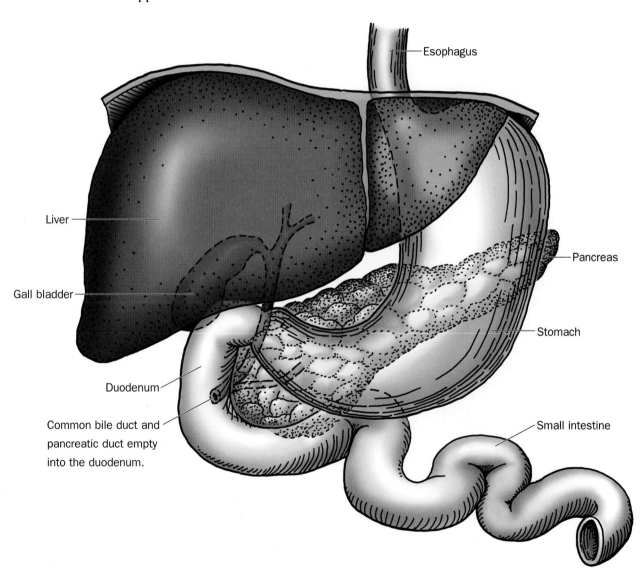

Liver

Gall bladder

Duodenum

Common bile duct and
pancreatic duct empty
into the duodenum.

Esophagus

Pancreas

Stomach

Small intestine

Figure 6.2 *The accessory organs of digestion*

Diffusion and Active Transport: Getting From Here to There

A lot of traveling goes on inside your body. Your blood is carrying nutrients and oxygen to every cell. But your cells are surrounded by membranes, and your blood is confined within blood vessels.

How do nutrients get past these barriers? There are two ways to do it: just go with the flow or hitch a ride.

Diffusion: Going With the Flow

What happens when you put a tea bag in a cup of hot water? At first, the tea is concentrated in one spot. But it soon spreads out. You can tell by the color change. The "wall" of the tea bag, however, keeps the tea leaves inside.

The particles that you taste when you sip the tea are so small they pass through the bag. The process that moves the tea particles and the water back and forth through the tea bag is called diffusion.

Inside your body, diffusion works much the same way. Digested nutrients dissolve and move through the walls of your small intestine and blood vessels. The blood then carries the nutrients throughout your body.

Diffusion that does not require any energy from cells is sometimes called passive transport.

Active Transport: Hitching a Ride

Your cell membranes let some things get through by passive transport. In that case, they are working in the same way that the tea bag does.

For some nutrients, getting through a cell's protective membrane is tougher. The membranes are choosy about what gets through and what does not. Because of this, they are sometimes referred to as "selectively permeable" or "semipermeable." In some cases, it takes energy from the cells to move nutrients from one side of the membrane to the other. The process that allows this to happen is called active transport.

If it weren't for diffusion, tea bags would be useless!

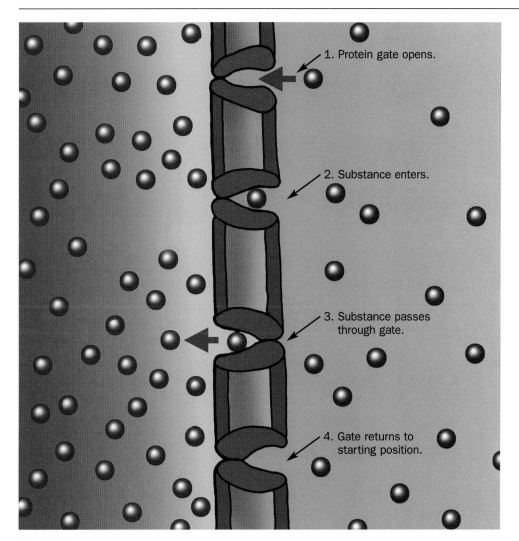

1. Protein gate opens.

2. Substance enters.

3. Substance passes through gate.

4. Gate returns to starting position.

Model of active transport

During active transport in the human body, substances use energy from the cell to move through the cell membrane. The energy comes from a substance called ATP (adenosine triphosphate), which is manufactured inside your cells. When a substance arrives at the cell membrane, a protein molecule grabs it, and ATP splits to release its stored energy. The energy changes the shape of the protein, which now acts as a gate in the cell membrane. When the gate (called a carrier) opens, the substance moves into (or out of) the cell. This uses up the energy the ATP has supplied to the carrier, and the gate closes.

Diffusion helps substances move through fluids and across cell membranes inside your body. When diffusion won't work, active transport may come to the rescue. These processes keep traffic moving smoothly to and from every cell in your body. □

SPIES

THE LONG AND WINDING TUBE

Peppi and Bollo are gradually moving downward as the contractions of the stomach continue. Looking down, they see a round black hole. The hole suddenly grows larger.

Whoosh! Peppi and Bollo pass through the opening, which quickly closes.

"Another sphincter," says Peppi. "Those muscles are powerful. That one was the pyloric sphincter. It's the gateway to the next important part of the digestive system, the small intestine."

"This is the duodenum," says Peppi. "It's a special name for the first 25 centimeters of the small intestine. It's hard to see from this perspective, but the small intestine is about 7 meters long."

"You've got to be kidding," says Bollo. "How can it all fit into that small space? And why do they call it the 'small' intestine if it's so long?"

"It fits because it's folded and tucked away," replies Peppi. "Even though it is long, it is probably called 'small' because of its diameter—only about 2.5 centimeters. Let's go along for the ride and see what happens."

A Change of Atmosphere

Peppi and Bollo are assaulted on all sides by spurts of liquid. One spurt comes from the direction of the pancreas.

"Pancreatic juice," Peppi explains, "is packed with enzymes that help digest carbohydrates, proteins, and fats. Each day, the pancreas secretes about 1.5 liters of juice.

"As food enters the duodenum, the gall bladder swings into action, contracting and pumping out greenish-yellow liquid bile. It's been stored there since it was manufactured in the liver. The liver, the largest organ in the body, performs many other functions as well."

"The pancreatic juice is different from the juices in the stomach," says Bollo. "It's not acidic. What's going on?"

"The juice from the pancreas is not acidic," says Peppi. "In fact, it's just the opposite: It is alkaline, or basic. It neutralizes the acid from the stomach and then starts off on some work of its own. Because digesting food is a big job, there's still more specialization.

"Foods get customized treatment at this point.

"Proteins, which were partially digested in the stomach, are acted on by intestinal and pancreatic juices. Eventually, they break down into amino acids.

"Carbohydrates, already well on their way to being digested, complete the change to simple sugars as a result of interaction with intestinal and pancreatic juices.

"Fats get a big dose of bile, courtesy of the liver and gall bladder. Bile works like dishwashing detergent. It breaks large fat droplets into smaller ones so they can mix more easily with the juices from the small intestine and pancreas. Fat eventually gets broken down into fatty acids and glycerol."

"More specialization!" says Bollo.

Peppi helps Bollo in a squeeze. It's the sphincter muscle that's causing his problem!

"Right!" says Peppi. "Both pancreatic and intestinal juices contain enzymes. Remember that enzymes are *specific*. What does that mean?"

"Each enzyme can digest only one type of nutrient," says Bollo.

"Right you are!" says Peppi. "At some point during the pizza's journey through the small intestine, digestion is complete. The pizza has been transformed into a soupy mixture of sugars, amino acids, fatty acids, vitamins, and minerals. The particles are now simple enough to be absorbed through the lining of the small intestine."

"How does this stuff actually get out into the body?" says Bollo.

"We're going to find out," says Peppi. "But first, let's take a break. The pace of activity in the small intestine has worn me out." □

Surface Area and Absorption

Lighting a campfire takes patience, and you have to start small. Why might that be?

INTRODUCTION

Have you ever built a campfire? Did you start with logs or twigs? Even though you will eventually need logs to keep the fire burning, you have to use twigs to get the campfire started.

Keep that idea in mind as you perform the inquiries in this lesson. Just as in Lesson 6, you will be investigating the small intestine. The topic you'll be exploring this time, however, is surface area. You'll begin your inquiry with a clay cube. You will measure it, calculate its surface area, and then explore how to increase its surface area. Next, you will build a model of a cross-section of the small intestine and calculate the surface area of your model. By the end of this lesson, you should be able to identify more reasons why the small intestine is such an efficient digestive organ.

OBJECTIVES FOR THIS LESSON

Calculate the surface area of a cube and a rectangular solid.

Explore how to increase the surface area of a cube while working within constraints.

Build a model of a cross-section of the small intestine.

Calculate and compare the surface areas of folded and unfolded strips of tickets.

Discuss the relationship between surface area and absorption in the small intestine.

Read about what happens to undigested food and water in the large intestine.

Getting Started

1. Review the Introduction to this lesson with your teacher.

2. With the class, discuss how you would calculate the surface area of the top of your desk. Then discuss how to calculate the *total* surface area of a cube. Finally, discuss how to calculate the total surface area of a rectangular solid. In your science notebook, record the formulas and any notes that might help you remember how to perform these calculations.

Inquiry 7.1
Increasing the Surface Area of a Clay Cube

PROCEDURE

1. Have a volunteer pick up a tote tray for your group.

2. Remove a lump of clay from the tote tray. Run your fingers over the entire piece of clay.

   How would you define the "surface area" of the clay?

3. Working with your partner, use the ruler to form the clay into a cube that measures 2 cm on each side.

MATERIALS FOR
LESSON 7

For you
1 copy of Student Sheet 7.1: Study Guide—The Digestive System

For your group
1 tote tray
2 30-cm rulers
2 lumps of clay in a plastic bag
2 plastic knives
2 30-ticket strips, fanfolded
2 12-ticket strips, unfolded
Masking tape

4. Using the formula you discussed in the "Getting Started" section, calculate the surface area of the cube. Record it in your science notebook.

5. With your partner, find a way to increase the surface area of the cube. Your method must meet two guidelines:

A. You must not greatly increase the total volume occupied by the clay. (In other words, if the clay came in a small box, it must still fit into that box once you have increased its surface area.)

B. You must still be able to calculate the total surface area of the cube.

6. Diagram your method in your science notebook. Defend your answer mathematically by calculating the new surface area and comparing it with the original area.

7. Share your method and results with the rest of the class.

8. Roll the clay into a ball and put it in the bag.

Inquiry 7.2
Modeling the Inside Surface of the Small Intestine

PROCEDURE

1. Working with your partner, form a cylinder with the 12-ticket strip. Join the ends of the strip with a small piece of masking tape.

2. Assuming that one side of each two-part ticket has a surface area of approximately 12.5 square centimeters (cm²), what

would be the total surface area of the inside of the cylinder? Record this in your science notebook.

3. Fanfold the strip of 30 tickets. Place the fanfolded strip of tickets on edge inside the circle of tickets to form an inner layer. Use a small piece of masking tape to join the ends of the folded strip.

4. You now have a model showing the folds in the walls of the small intestine. Compare your model, which is illustrated in Figure 7.1, with the illustration of a cross-sectional view of the small intestine that appears in Figure 7.2.

Figure 7.1 *Model of a cross-section of the small intestine*

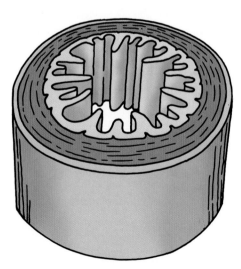

Figure 7.2 *Cross-section of the small intestine*

5. Calculate the surface area of the inside of the folded strip of tickets and record it in your science notebook.

6. Compare the surface areas of the folded and unfolded tickets and express the relationship between the surface areas as a ratio (for example, 5-to-1). Record your answer in your science notebook.

REFLECTING ON WHAT YOU'VE DONE

1. Discuss your method of increasing the surface area of the clay. Explain how you proved that the surface area increased.

2. Share your results from Inquiry 7.2. Discuss the advantages of the enormous surface area of the small intestine. Do you think that other human organs might need a large surface area? Which ones? Record your ideas in your science notebook.

3. Your teacher will ask two questions about surface area. Discuss your answers with the class.

4. Many tools and instruments we use in daily life are designed with folds. How many can you can you think of? What are the advantages of the folds? Consider the accordion in the photo to the right. Discuss this with the class.

5. On the basis of what you read in "Spies: Leftovers," summarize what happens to water and undigested wastes in the large intestine. Record these ideas in your science notebook. Be sure to mention the conditions that may occur when food moves through the large intestine too quickly and too slowly.

CORBIS/TODD GIPSTEIN

The bellows of this accordion have a surface that is similar to the folded surface of the small intestine. This accordion player would not be able to carry his instrument around as easily if it did not have folds.

SURFACE AREA:
Your Intestine Isn't Small at All

What's the best way to get the most out of the nutrients in your food? It is to make sure that most of those nutrients get into the right places in your body.

Your digestive system has some special features to make this happen. First, it has ways of making sure the surface area of the foods you eat is as large as possible. Second, your digestive system packs as large a surface area as possible into a limited space.

As soon as you begin to chew your food, you start to increase its surface area. Your teeth break the food into smaller and smaller pieces. The surface area of the food continues to increase, allowing the digestive enzymes greater access to them. When the food gets to your stomach, it is broken into even smaller pieces by gastric juices and muscular action. Then the food mixture moves into your small intestine.

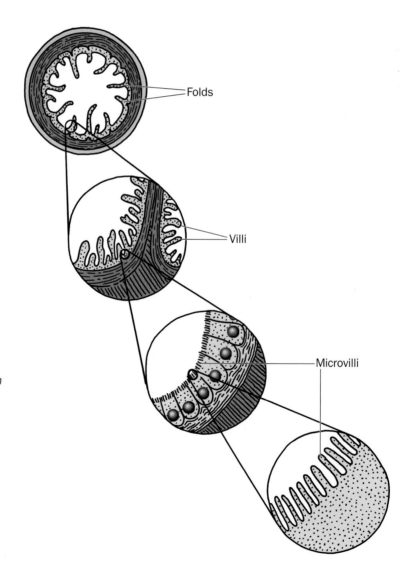

Folds

Villi

Microvilli

Lining of the small intestine, shown with increasing magnification

The small intestine is a hollow tube about 7.0 meters long and 2.5 to 5.0 centimeters in diameter. The first 25 centimeters of the small intestine is known as the duodenum. That's where most digestion is completed. As the digested carbohydrates, proteins, and fats move through the remainder of the small intestine, they are absorbed through the intestinal wall into your bloodstream. If its inside wall were smooth, your small intestine would have an internal surface area that is about the same as about the top of a dinner table. That's a lot, but it's not enough to let all the nutrients you need pass into your bloodstream.

So maybe it's not surprising to learn that the inside of your small intestine isn't smooth. Instead, it is packed with wrinkles and folds. The surface of these folds is covered with tiny projections called villi. The villi are covered with even tinier projections known as microvilli.

The folds, villi, and microvilli may be compared with the fluffy threads of a bath towel. Just like the threads of a terry cloth towel absorb water better than a flat cloth, the villi allow your intestine to absorb more nutrients than a smooth surface would.

The folds, villi, and microvilli increase the internal surface area of the intestine by about 600 times. In fact, scientists estimate that if you could flatten out the surface of your small intestine, it would almost cover a football field. That's a lot bigger than a table top. ☐

© G. SHIH-R. KESSEL/VISUALS UNLIMITED

Microscopic, fingerlike villi line the walls of the small intestine and increase its surface area (SEM × 140).

LEFTOVERS

Peppi and Bollo have now seen more than their share of villi and microvilli. They've traveled nearly the entire 7 meters of the small intestine.

Just when they're about ready to stop for a break, they see an opening ahead.

"The large intestine, right?" says Bollo.

"That's one word for it," Peppi replies. "It's also called the colon."

"What's that opening?" Bollo asks, looking to his right.

"That's the entrance to the appendix," says Peppi. "Scientists on planet Earth believe that the appendix once helped humans digest the cellulose in plant matter.

"The only time humans ever hear about the appendix today is when it causes trouble—appendicitis. When the appendix becomes blocked, it may become swollen and infected by bacteria. This causes pain. If the appendix bursts, it releases bacteria into the abdomen. This can be serious. Appendicitis is treated by surgery. Since the appendix no longer has a function, a human wouldn't miss it."

Peppi and Bollo have nearly completed their trip through the villi. Next stop: the large intestine!

Peppi and Bollo in the large intestine. Peristalsis keeps things moving here!

Peppi and Bollo continue their trip. First they go up. Then they turn sharply. As they travel, the amount of water decreases. The undigested waste clumps together.

"In some ways, the colon is like the body's storage tank," explains Peppi. "It doesn't produce enzymes. It doesn't need to, because the food is already digested. The leftover food, now undigested waste, spends time here before it is eliminated through the anus.

"The name for the material that passes through the colon is 'feces.' Feces are composed of water, undigested food, mucus, dead cells, and bacteria.

"Even though the colon has no role in digestion, it has an important part in keeping the body in balance. That's because of its role in the absorption of water and minerals. Water and minerals pass through the walls of the large intestine back into the bloodstream."

"Nothing very exciting about that," says Bollo. "I think we know a lot about absorption already."

"You're right. But things can go wrong here, even though no digestion is taking place. If the feces move through the large intestine too quickly, there's not enough time for water to be absorbed. The result? Diarrhea. Diarrhea is often accompanied by cramps, which are strong contractions of the wall of the large intestine. Diarrhea can be dangerous, especially in an infant or a young child. If humans lose too much water, they become dehydrated.

"Sometimes the opposite situation occurs. The feces move through the colon too slowly. Too much water is absorbed. The feces become larger, harder, and drier. The result is constipation. One way to prevent constipation is to eat foods with high fiber content. If that does not relieve constipation, humans sometimes take a laxative. A laxative is a medication that causes the peristaltic contractions to increase and to move wastes through the colon more quickly."

"Yuk. It smells awful in here," says Bollo.

"That's gas. It's released when bacteria in the large intestine start to feast on undigested food.

"If too much gas builds up, it causes pain. One way to relieve it is to let the gas pass out through the anus. But that isn't polite in most human societies!"

Pretty soon the action comes to a halt. The feces collect at the end of the colon, in a place called the rectum.

Suddenly the wall of the colon starts to contract powerfully. Peppi and Bollo try to hold on. Out they go into a brilliant white bowl with crystal-clear water!

"Let's get out of here before it's too late!" says Bollo. They grab the top rim of the bowl. Directly below, they see a powerful whirlpool. Water and feces disappear.

"We made it!" says Peppi. "Back to civilization at last!"

Summing Up

"Let's review what we've learned," says Peppi, as they dry off on the top of the toilet tank.

"First, the digestive system works like a food-processing factory. Its job is to change raw materials into a form the body can use to grow and maintain health. The body does this through mechanical and chemical actions. Now what else have you learned, Bollo?"

"Well, the digestive system is specialized. Different organs have different functions," says Bollo.

"You're right. A pretty impressive system, wouldn't you agree? We have learned quite a bit, but we're far from finished. There's another exciting journey just ahead." □

The spies' first journey is complete, but more adventures lie ahead.

The Digestive System— An Assessment

For the first part of your assessment, you will perform a chemical test.

INTRODUCTION

Congratulations! You have now completed your investigation of the human digestive system. This assessment will give you an opportunity to apply your new skills and knowledge. It has two parts. In Part 1, you will use your laboratory skills to perform a chemical test. Part 2 of the assessment has two sections. First, you will answer a series of questions about the main ideas covered in Lessons 1 through 7. You'll answer these questions on your own. Then, with your group, you will revise your human body systems poster.

Although you have reached the end of your exploration of the digestive system, there's still much more to come. Lesson 9 introduces you to the Anchor Activity, a research project that you and a partner will work on during the rest of this module. Lesson 10 marks the beginning of your investigation of the respiratory and circulatory systems. Peppi and Bollo will be waiting for you!

OBJECTIVES FOR THIS LESSON

Design and perform an inquiry to determine which of two containers of starch solution contains an enzyme produced in the human mouth.

Respond to a series of questions and statements based on the main ideas and skills covered in Lessons 1 through 7.

Revise your group's human body systems poster.

Inquiry 8.1
Which Solution Has the Enzyme?

PROCEDURE FOR PERIOD 1

1. Have one member of your group pick up your supplies.

2. Discuss the Safety Tips that you followed when you performed chemical tests. See how many tips you can remember.

3. With the class, review the instructions for Part 1 of this assessment. Listen to your teacher's directions for conducting the inquiry, recording and turning in your procedure and results, and cleaning up.

4. Review Table 8.1: Inquiry Scoring Rubric and Table 8.2: Inquiry Checklist on page 63. Your teacher will use these forms or similar ones to evaluate your performance on Part 1 of the assessment.

5. Begin to design your inquiry. Answer items A, B, and C on Student Sheet 8.1a before you begin your investigation.

6. Turn in Student Sheet 8.1a before you leave class.

MATERIALS FOR LESSON 8

For you
- 1 copy of Student Sheet 8.1a: Part 1—Which Solution Has the Enzyme?
- 1 copy of Student Sheet 8.1b: Part 2A: Selected-Response Items
- 1 copy of Student Sheet 8.1c: Part 2A: Selected-Response Items—Answer Sheet
- 1 pair of safety goggles

For your group
- 1 tote tray
- 2 dropper bottles of Benedict's solution
- 2 dropper bottles of Lugol solution
- 4 containers of starch solution (2 labeled "A" and 2 labeled "B")
- 2 6-cup lab trays
- 2 medium test tubes
- 1 test tube rack
- 2 test tube clamps
- 2 plastic pipettes
- 1 human body systems poster
- 1 roll of clear tape
- 1 black marker
- 1 pair of scissors
- 1 eraser

PROCEDURE FOR PERIOD 2

1. Read the questions on Student Sheet 8.1b. Enter your answers on Student Sheet 8.1c.

2. When you have finished, turn in both student sheets.

3. When everyone in your group has turned in his or her student sheet, get to work on your poster. Thinking about everything you now know about the digestive processes, look it over carefully. Then do the following:

   A. Decide whether anything needs to be corrected or changed. Use a spare summary box if you want to rewrite one of your descriptions.

   B. If any organs are in the wrong place, move them to the correct location.

   C. Tape each summary next to the appropriate organ.

   D. Give the poster to your teacher

4. Read "Feeling Good, Looking Better!" at the end of this lesson

REFLECTING ON WHAT YOU'VE DONE

1. Share in a class discussion about the results of your inquiry and your responses to the items in Part 2A of your assessment. Take a look at the other groups' posters. Discuss any questions that you have.

2. With the class, discuss "Looking Good, Feeling Better!" Suppose, for example, that you thought that one of your friends or relatives had an eating disorder. What could you do to help that person?

3. Take another look at the questions you listed in your science notebook at the end of Lesson 2. Discuss any questions that were not covered in this part of the module.

Table 8.1 Inquiry Scoring Rubric

Activity	Exemplary	Satisfactory	Needs Improvement	No Attempt
Designing the inquiry				
Setting up the materials and carrying out the inquiry				
Recording and representing data				
Interpreting data				
Working collaboratively				

Table 8.2 Inquiry Checklist

Skill	Score
State the question in an appropriate manner.	
Generate a list of materials that you will need for the inquiry.	
Organize the procedure you will follow into logical steps.	
Use an appropriate control for your experiment.	
Control all the variables except the one you are testing.	
Construct an appropriate data table.	
Conduct enough trials to obtain sufficient data.	
Collect data that fall within reasonable expectations.	
Display your data using an appropriate form of graph or table (if applicable).	
Interpret your data in a logical way.	

LOOKING GOOD, FEELING BETTER!

Jenny's mirror did not reflect reality.

Jenny was 14. When she looked in the mirror, she thought she saw a stranger. She had grown more than 2 inches in the past year. She'd also gained weight.

Jenny was 5′6″ tall, and she weighed 110 pounds. "I'm too fat. I've got to diet," she said.

Jenny knew about nutrition. She began by cutting down on fat. No more french fries! She also stopped eating chocolate.

She soon lost a few pounds, but she wanted to lose more. She ate less and less and began to exercise more and more.

By the end of a month, Jenny had lost 15 pounds. But she didn't feel good. She was pale. She was always tired, and she felt jumpy. Her swimming coach was concerned. So were her parents. Her mother took her to the doctor.

The Doctor's Advice

"You don't need to lose weight, Jenny," said Dr. Parsons. "In fact, you need to gain back some of the weight you've been losing. From what you've told me, you've been eating under 1000 calories a day. A girl of your age should be eating twice that many calories. It's okay to cut back on fat, but your diet is not giving you the all the nutrients you need. Let's design a well-balanced diet that will help you gain back some of that weight and begin to feel good again.

"I also want you to read this information about eating disorders. Now don't worry. I don't think you have any of the problems described in this booklet. But you should know about them."

What Are Eating Disorders?

Jenny and her mom read the information from Dr. Parsons. One eating disorder, overeating, was easy to understand. If you regularly take in more calories than you use up, you gain weight. People whose weight exceeds 20 percent of their recommended

weight are described as obese. This means they have extra body fat. Obesity is a health risk.

Another eating disorder is bulimia. Almost all of the people with this disorder are women. These women are usually of normal weight. They eat large amounts of food; then, they make themselves vomit to prevent weight gain. Some take laxatives for the same reason. Bulimia is often called the binge-and-purge syndrome. Over the long term, it can seriously damage the body.

Then Jenny read about an eating disorder called anorexia. It is most common in girls and women between the ages of 15 and 30, but it also occurs in boys and men.

Anorexia may begin with a desire to lose a few pounds. But it soon takes over every aspect of a person's life. People with anorexia are obsessed with food. They think about food almost constantly, but they do everything they can to avoid eating.

They exercise frantically. They pride themselves on their willpower. In reality, they are starving themselves.

People with anorexia almost always feel cold, because they have little body fat. Their skin becomes dry and cracked. As their muscle tissue deteriorates, they may develop cramps. Girls with anorexia stop menstruating. Their hair may fall out.

Anorexia is a serious disease. If untreated, it can damage the heart, kidneys, other organs, and bones. Sometimes it is fatal.

Back to Dr. Parsons

Two weeks later, Jenny and her mother went back to Dr. Parsons. Jenny had gained 3 pounds. "That's great," said Dr. Parsons. "Now let's talk about the information you read."

"Anorexia sounds serious," said Jenny. "What causes it?"

"Anorexia is caused by a combination of many pressures, both internal and external. First of all, there's pressure in our society today to be extremely

thin. Many advertisements and photos of fashion models convey that message.

"Some people have other reasons for wanting to be thin," Dr. Parsons continued. "For example, some athletes think that being thin helps them be more competitive. In fact, the opposite is often true.

"Some patients say the disease began at a time when they were upset—after a death in the family or their parents' divorce.

"These are things that researchers are exploring. What's important is that everyone who has this disorder gets the help they need. The first step is to visit a doctor."

"I'm glad we had this talk," said Jenny. "It's made me realize that I need to eat sensibly. I still want to look good. But that doesn't mean looking like a model in a fashion magazine. What's most important is to be healthy and strong. I want to look good, but I also want to feel good—*and* feel good about myself. And now I do!" □

The Respiratory and Circulatory Systems

Anchor Activity— Diseases and Health Careers

The Internet will be a helpful resource for your Anchor Activity research.

INTRODUCTION

Diseases have had a strong impact on human history. For example, a disease called the bubonic plague killed millions of people in the Middle Ages.

Because of scientific advances, some diseases that used to cause widespread suffering and death have been nearly wiped out. The severity of other diseases has been greatly reduced. In your parents' lifetime, for example, a vaccine has eliminated smallpox from the world. Tuberculosis, a serious lung disease, can often be cured. When public health measures are in place, the spread of this disease can be controlled.

New diseases are always emerging, however. For example, acquired immune deficiency syndrome (AIDS) was unknown until the last quarter of the 20th century.

OBJECTIVES FOR THE ANCHOR ACTIVITY

Research a disease that has a major effect on the human body,

or—

Explore a health career that involves disease research, diagnosis, treatment, or prevention.

Use a variety of tools and techniques to gather, interpret, and display your data in poster format.

Make an oral presentation to the class summarizing your research.

What are diseases? What causes them?
How are they spread? How do they affect the
human body? How are they diagnosed, treated,
prevented, and cured? What role do health
professionals—doctors, nurses, researchers,
technicians, health educators, and others—have
in keeping people healthy and disease-free?

The Anchor Activity that begins in this lesson
will help you and your classmates answer questions such as these. You and your partner will
choose one disease or health career to research
in detail. Over the next few weeks, you will work
on your research and organize your findings.
Finally, you'll share your findings with the class
in a poster presentation.

**MATERIALS FOR
LESSON 9**

For you
1 copy of Student
Sheet 9.1a: Anchor
Activity Time Line
1 copy of Student
Sheet 9.1b: Anchor
Activity Scoring
Rubrics

**For you and
your partner**
1 sheet of poster
paper
Glue or clear tape
Scissors

Getting Started

1. Discuss with the class the answers to the
questions following the reader entitled
"Disease: What's Gotten Into You?" in
this lesson.

2. With your group, discuss the following
questions. Record your answers in your
science notebook.

   *A. How many health careers can you
   think of? List them.*

   *B. What do you think someone should
   know about a health career before deciding to enter it?*

3. Share your responses with the class.

Anchor Activity

PROCEDURE

1. Listen and follow along on Student Sheets 9.1a and 9.1b as your teacher introduces the Anchor Activity. With your partner, you will research a disease or health career, organize your materials, and create a poster. Then you will summarize your findings in a 2-minute oral presentation to the class. Your teacher will give you the deadline for completing your poster. Your teacher will also set other deadlines for completing the preliminary stages of your work. These deadlines appear in the left-hand column (labeled "Date Due") on Student Sheet 9.1a.

2. You and your partner may select the topic for your research.

A. If you decide to research a disease, you may consider the following topics:

Addison's disease	Huntington's disease
AIDS	Hydrophobia (rabies)
Alzheimer's disease	Hypoglycemia
Anthrax	Jaundice
Arthritis	Leukemia
Asthma	Lou Gehrig's disease
Attention deficit disorder	Lupus
Cancer	Lyme disease
Chickenpox	Malaria
Chronic bronchitis	Meningitis
Chronic fatigue syndrome	Multiple sclerosis
Cirrhosis of the liver	Mumps
Colitis	Myocarditis
Congenital heart disease	Osgood-Schlatter disease
Crohn's disease	Parkinson's disease
Degenerative joint disease	Peptic ulcer
Diabetes	Pneumonia
Emphysema	Psoriasis
Endocarditis	Rheumatic heart disease
Epilepsy	Sleep apnea
Glaucoma	Smallpox
Infectious mono- nucleosis	Strep throat
Influenza	Stroke
German measles	Tetanus (lockjaw)
Hepatitis	Tuberculosis
Hodgkin's disease	Typhoid fever
	Whooping cough
	Yellow fever

B. If you decide to research a health career, you may consider the following ideas:

Allopathic physician	Industrial hygienist
Anesthesiologist	Licensed practical nurse
Audiologist	
Biomedical engineer	Nurse practitioner
Chiropractor	Occupational therapist
Clinical laboratory technician	Operating room technician
Dental assistant	Ophthalmologist
Dental laboratory technician	Optician
	Osteopathic physician
Dentist	Pediatrician
Dialysis technician	Pharmacist
Dietitian	Physical therapist
Electrocardiograph technician	Podiatrist
	Prosthetist
Emergency medical technician	Psychiatrist
	Psychologist
Environmental health specialist	Radiation therapist
	Registered nurse
Epidemiologist	Research scientist
Health care adminis- trator	Respiratory therapist
	Speech pathologist
Health educator	X-ray technician

3. Once you and your partner have picked your topic, follow the directions in the first block on Student Sheet 9.1a: Anchor Activity Time Line to record your idea. Write it down on a sheet of paper and give it to your teacher. You and your partner will have to divide the responsibilities and do your research individually. Once you have both completed your research, you will work together to organize your material.

4. Your presentation should include information on each of the following areas (under "4A. Disease" or "4B. Health Career"):

A. Disease

- Cause of the disease. If caused by a pathogen, include an illustration of the pathogen.
- Scientists who discovered the cause of the disease. Who? When? Where? How?
- Symptoms of the disease.
- Effects of the disease. What short- and long-term effects can this disease have

on the body? (Be sure to relate what you have learned about the affected body systems.)

- Prevention. Can the disease be prevented? If so, how? What scientists have made contributions to preventing this disease?
- Treatment and cure. If no cure has been found, how close are scientists to developing one? What scientists have made contributions to the search for a cure?
- Technology implications. How have advances in technology affected the way in which this disease is diagnosed and treated?
- Self-treatment. Can a person with the disease control it by changing some of his or her daily habits?

B. Health Career

- Job title. Include a photo or an illustration of a person who has this job title.
- Job description. Describe in detail the duties of a person in this career.
- Educational requirements.
- Salary. What is the salary range? What is the potential for advancement?
- Job outlook. Describe the need for this career in the future.
- Lifestyle implications. How does this career affect one's personal life? (For example, how many hours a day does the person work? Does the person have to work nights? Does the job involve contact with people, or does the person usually work alone?)
- Technology. How has technology influenced this career over time? (For example, what forms of technology have led to improvements in diagnosing problems or treating patients?)

5. Use a variety of references to create your poster. Sources of information might include your school media center, the World Wide Web, public libraries, personal interviews, CD-ROMs, DVDs, magazines, newspapers, encyclopedias, books, and videotapes. Your final product must include references from the following:

- At least two Internet or CD-ROM/DVD sources
- At least two print sources

If you are researching a career, consider doing an interview with someone working in the career you have chosen. Take along a camera to take a photo of the person you are interviewing.

6. List your references in a bibliography. Prepare the bibliography on a separate sheet of paper and turn it in with your poster. Your teacher will give you information on how to format your references.

7. Organize your poster so that the required information described in Step 4A or 4B of the Procedure is clearly visible. Make your poster as creative and interesting as you can.

8. Plan your oral presentation. It should contain highlights from your poster. Be sure to practice your presentation beforehand.

A. If you researched a disease, include the name of the disease; its cause, treatment, prevention, and cure (when applicable); and the scientists recognized for their work with the disease.

B. If you researched a health career, include the name of the career, its educational requirements, and a brief description of the duties of a person in this career.

9. When all the presentations have concluded, your teacher will display all the posters and give you time to look at them.

Disease: What's Gotten Into You?

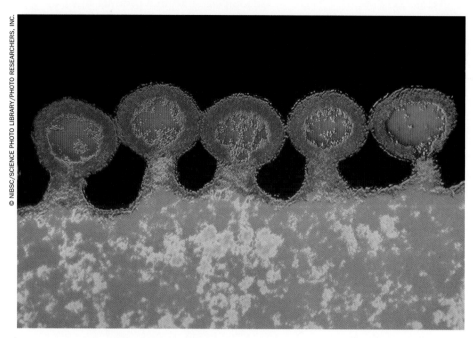

An electron micrograph of the human immunodeficiency virus (HIV), which causes acquired immune deficiency syndrome (AIDS). The viruses are budding from a host white blood cell. When they do this, they damage the cell membrane. The cell dies and the body's immune system is weakened (magnification ×72,000).

Bacteria: Unwelcome Guests

Bacteria are single-celled organisms. They are so small that you can't see the individual cells without a microscope. When they find the right environment—a place like your body, where it's warm and moist and there's plenty of food—they reproduce readily.

Most bacteria are harmless. Some are even helpful. But there are also about 100 types of bacteria that are dangerous, and they are a major cause of human illness.

Bacteria can get into your body when you breathe. There are also bacteria on the

Why do you get sick? Most of the time, the answer is simple. You've been invaded by aliens!

These germs—tiny organisms that invade your body and cause disease—are known as "pathogens." (The Greek word for "sickness" is *pathos*.) The two most common kinds of pathogens, bacteria and viruses, are responsible for many human diseases.

Other things cause disease, too. If your body doesn't produce the right amounts of the chemicals it needs, you might get a disease like diabetes. You might get sick if you're exposed to a harmful substance in air, food, or water. Other diseases, such as sickle-cell anemia, are inherited.

There are many ways to get sick, but your body's immune system has powerful weapons to fight those alien invaders. And when

you do get sick, there are medicines that can often help you get well.

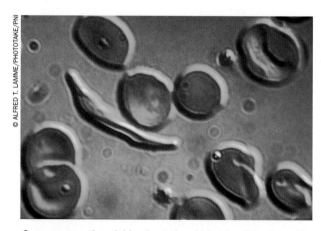

Can you see the sickle-shaped red blood cell in the middle left section of this photograph? The distinctive shape interferes with the cell's ability to carry oxygen.

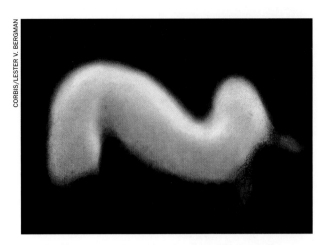

The Spirillum minus *bacterium is named for its spiral shape. It causes spirillar fever, a form of rat-bite fever.*

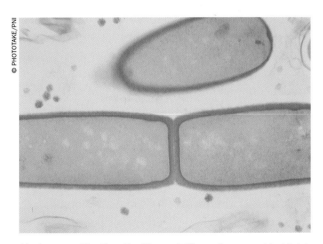

Under magnification, Bacillus subtilis *cells resemble bluish, rectangular boxes. These bacteria often occur in chains.*

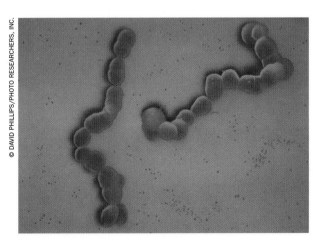

Streptococcus bacteria are spherical. They often occur in chains and resemble a pearl necklace.

food you eat. Bacteria can invade your body if your skin is cut or scratched. Bacteria can damage cells directly, or they can produce substances called toxins, which are poisons.

Bacteria are specialists. Some prefer to live in your digestive system. They may cause abdominal cramps, diarrhea, and vomiting. When your doctor says you've got "strep throat," he or she usually means that a colony of streptococcus bacteria are living in your throat.

Viruses: Tiny Packets of Trouble

Viruses are even smaller than bacteria. They must invade your cells in order to survive and reproduce. Like bacteria, viruses can get into your body through the air you breathe or the food you eat. Some viruses, as well as some bacteria, are spread through sexual contact.

Once the viruses have invaded your body, they attach themselves to your cells and penetrate the cell membranes. Then they trick normal cells into producing more virus particles, which rupture the cell membranes and move on to infect other cells.

Viruses cause many diseases. Some of these diseases are as simple as the common cold. But other viral diseases can cripple or kill. Hepatitis, rabies, polio, and AIDS are all caused by viruses.

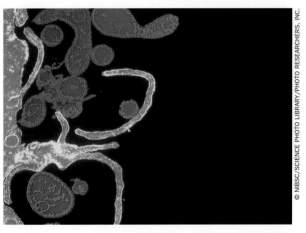

Measles viruses budding from the surface of an infected cell (magnification ×14,400).

Prevention: Protecting Your Territory

"I think you ought to put a bandage on that scratch."

"Cover your mouth when you sneeze."

Simple things can help protect your body from pathogens or help prevent you from infecting other people. Your body has its own built-in protection, too. Your skin forms a seal to keep most things out. The mucus in your mouth and nose can trap invaders and then destroy them with enzymes. The gastric juice in your stomach can kill germs.

When a bacterium

Dr. Edward Jenner vaccinates 8-year-old James Phipps against smallpox.

or virus gets into your body, your immune system goes to work.

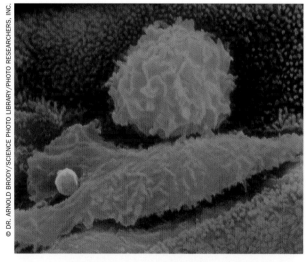

Two macrophages in a human lung. Macrophages can clear the lungs of dust, pollen, and bacteria; however, some pollutants can destroy the macrophages. This can lead to pulmonary disease.

Special cells recognize it as foreign. These cells release chemicals that bring other defenders to work. Some are cells called macrophages (which means "big eaters") that surround and break apart the germs.

Other cells memorize important information about the invader. The memory cells circulate through your body to recruit more defenders and tell them what to look for. When the infection is gone, your body saves a chemical record—like a

chemical memory—in substances called antibodies. The next time the same pathogen gets into your body, the antibodies recognize it and destroy it before it makes you sick again.

The shots your doctor gave you before you started school contained pathogens that were too weak to make you sick. They tell your immune system how to identify dangerous germs.

This kind of medicine is called a vaccine. The first vaccine was invented in 1789 by an English doctor named Edward Jenner. Modern vaccines protect people from diseases like hepatitis, polio, and measles. Although the vaccines may prevent these diseases, they cannot cure them.

Treatment: Battling the Bad Guys

Many medicines do not actually cure disease. They just help you feel better while your body heals itself. In other words, they relieve the symptoms of disease.

For example, drugs such as aspirin lower your temperature if you have a fever.

Other medicines kill the bacteria that are making you sick. Antibiotics are a good example of this kind of medicine. In 1928, a Scottish doctor named Alexander Fleming discovered penicillin, which was the first antibiotic.

When penicillin first became available, it was called a wonder drug. Since then, scientists have developed many more antibiotics to treat diseases caused by bacteria. Doctors used to prescribe antibiotics to treat many diseases, but now doctors are more careful.

Why? Because bacteria are beginning to become resistant to antibiotics. Bacteria that are especially strong may not be killed by antibiotics. When these bacteria reproduce, they pass on their resistance to the next generation. If antibiotics kill only the weakest bacteria, the strongest types become more common. The antibiotics don't always work against these superpowerful bacteria.

Some doctors today will not prescribe an antibiotic to control a mild infection. That way, the bacteria population stays weak, and antibiotics will still work when they are really needed.

Penicillin, the first "wonder drug," is extracted from Penicillium *mold such as this.*

You can help, too. If your doctor tells you to take your medicine for a week, don't stop after 3 days, even if you feel better. That way, the antibiotic will kill as many of the bacteria as possible. ☐

QUESTIONS

1. What are three diseases caused by bacteria?
2. What are four diseases caused by viruses?
3. What are two ways in which your body deals with disease-causing microorganisms?
4. Why is it important to take all the antibiotic tablets that your doctor prescribed for you, even if you already feel better?

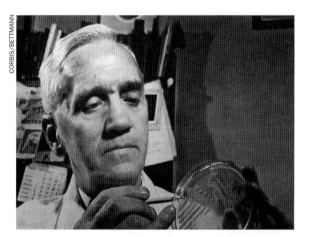

Dr. Alexander Fleming examines a culture dish containing Penicillium *mold. He is pictured in his laboratory at* St. Mary's Hospital in Paddington, Scotland.

Assessing Breathing Models

CORBIS/JEAN-YVES RUSZNIEWSKI; TEMPSPORT

Breath control is essential for swimmers, especially those who compete in meets.

INTRODUCTION

In the first part of this module, you explored how your body digests food and how nutrients are absorbed into the bloodstream. Food is one of the main ingredients your body needs to function. Another one is oxygen. How does your body take in oxygen? And then what happens?

To answer these questions, you need to know about the structure of a second human body system—the respiratory system. You also need to know how the process of breathing works.

You will begin this lesson by discussing what you already know about breathing. You'll examine a model that is often used to simulate how air enters and leaves the lungs. Then you will design and construct your own model of the breathing process. On the basis of what you know about breathing and the structure of the human respiratory system, you will discuss the strengths and limitations of each model.

OBJECTIVES FOR THIS LESSON

Review what you already know about the breathing process.

Observe a demonstration of the bell jar model of breathing.

Design, construct, and operate a model that simulates the breathing process.

Assess the strengths and limitations of two breathing models.

Recognize that all models have limitations.

Become familiar with the basic structure and functioning of the human respiratory system.

EXCUSE ME, PLEASE!

MATERIALS FOR LESSON 10

For you
1 copy of Student Sheet 10.1: Assessing the Syringe Model of Breathing

For your group
1 tote tray
2 syringes
2 balloons
2 400-mL beakers
Water (or access to a sink)

Sometimes you can't help it. You do something that your parents and teachers have told you isn't polite. You burp. Or you get the hiccups. Embarrassing!

What causes burps and hiccups? Let's take a look.

Burps: Doing Away With Extra Air

A burp is your body's way of getting rid of swallowed air. That air gets into your stomach by way of the esophagus, just as food does. But food is supposed to be in the stomach; air belongs in the lungs, right?

If you "swallow" too much air while you're eating, you'll start to feel uncomfortable. It's especially likely to happen if you're eating fast.

Once you've swallowed too much air, you'll get an unpleasant sense of fullness that is caused by the pressure of the air. Burping gets rid of the extra air and relieves the uncomfortable feeling.

When you drink a soda, you swallow carbon dioxide bubbles. That can lead to burping. Or maybe you've "burped" your baby brother or sister. Burping babies is necessary because they swallow air when they are

being fed. If they get too much gas, their stomachs expand. That hurts, and the baby usually starts to cry. By placing your baby sister on your shoulder and patting her gently on the back, you can help the air come back up.

Although burping is frowned on in our society, it's a compliment to the host in some other cultures. It's kind of like saying, "Thank you for an excellent meal. I enjoyed it!"

Hiccups: Things Get Out of Synch

Hiccups can be worse than burps. Once you get them, you think they'll never end!

Hiccups are a sign that your breathing is out of synch. To understand why hiccups happen, you have to know something about how the respiratory system works.

Respiration involves two main types of muscles—the rib, or intercostal, muscles and the diaphragm, which lies at the base of your lungs.

Normally, these two muscle groups work together like a well-synchronized machine. When you inhale, the intercostal muscles contract, and your rib cage moves up and out. The diaphragm moves down and flattens. This makes room for air in your lungs. A moment later, when you exhale, the diaphragm and intercostal muscles relax, forcing air out of the lungs.

Hiccups happen when the diaphragm contracts and pushes down at the wrong time, forcing air to move quickly past the vocal cords. Your brain says, "Wait a minute!" It sends a message to the tongue and the back of your throat to stop that air.

When air is forced across the vocal cords in the back of your mouth, the cords snap shut. You make a funny sound, called a hiccup.

People try many things to stop the hiccups. Some people eat a spoonful of sugar. Some breathe into a bag. Others believe the cure is to hold their breath and drink water.

The best thing to do is just to relax and try to breathe regularly. Soon things will get back in synch and the "hic-hic-hic" will stop.

Getting Started

1. Listen as student volunteers read "Excuse Me, Please," about burps and hiccups, in this lesson. Your teacher will use this reading selection to introduce you to the respiratory system, which is shown in Figure 10.1.

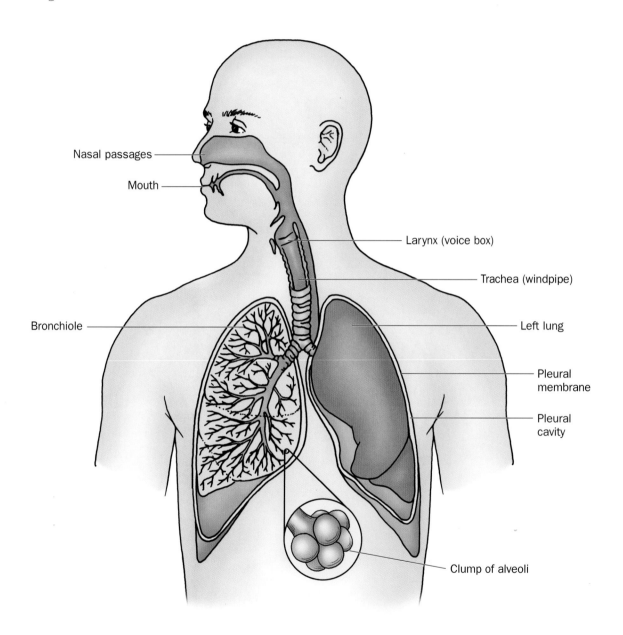

Nasal passages

Mouth

Larynx (voice box)

Trachea (windpipe)

Bronchiole

Left lung

Pleural membrane

Pleural cavity

Clump of alveoli

Figure 10.1 *The respiratory system*

2. Now it is time to focus on what happens when you breathe. Have one member of your group stand up, take a deep breath, and let it out slowly. Have the person repeat the procedure a few more times. In your science notebook, record everything you see happening.

3. Share your list with the class.

4. Listen as your teacher explains the breathing process and how gases are exchanged in the lungs. These actions are illustrated in Figures 10.2 and 10.3.

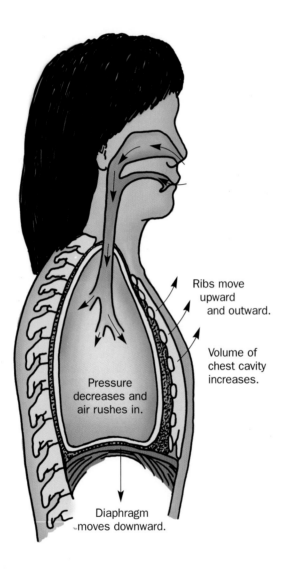

Ribs move upward and outward.

Volume of chest cavity increases.

Pressure decreases and air rushes in.

Diaphragm moves downward.

Inhalation

Ribs move downward and inward.

Volume of chest cavity decreases.

Pressure increases and air moves out.

Diaphragm moves upward.

Exhalation

Figure 10.2 *The breathing process*

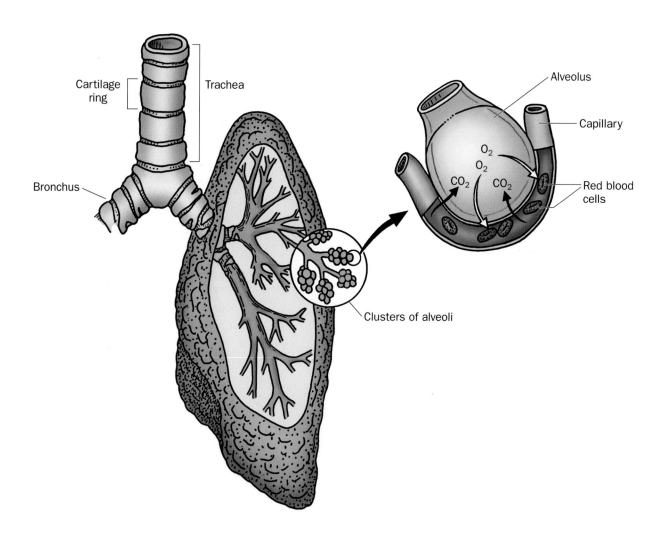

Figure 10.3 *Gas exchange in the lungs*

5. Now watch as your teacher demonstrates the bell jar model, which is often used to simulate the breathing process. The model is shown in Figure 10.4.

6. Turn to a blank page in your science notebook. At the top, write the words "The Bell Jar Model of Breathing." Draw a vertical line down the middle of the page. Label the left column "Strengths" and the right column "Limitations." On the basis of what you understand about the breathing process, make a list in each column. "Strengths" are features in the model that accurately show what happens when humans breathe. "Limitations" are characteristics of the model that fail to show (or do not show accurately) what actually happens during breathing.

7. Share your list with your class.

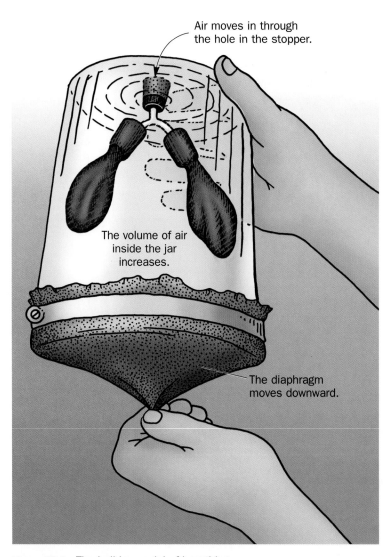

Air moves in through the hole in the stopper.

The volume of air inside the jar increases.

The diaphragm moves downward.

Figure 10.4 *The bell jar model of breathing*

Inquiry 10.1
Assessing the Syringe Model of Breathing

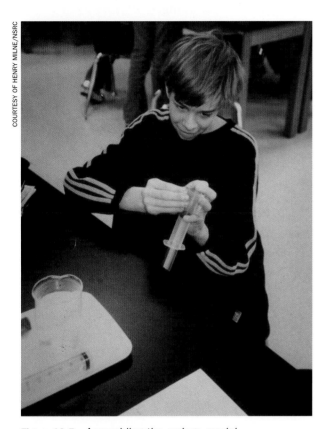

Figure 10.5 *Assembling the syringe model*

PROCEDURE

1. Have someone pick up your group's materials. Then watch as your teacher demonstrates how to operate the syringe.

2. You will work in pairs for this inquiry. Your challenge is to design and build a breathing model using a syringe, a balloon, and water (see Figure 10.5). Before you begin to construct your model, think about the following questions and answer them in your science notebook:

A. *What part of the syringe can be compared with the diaphragm? The mouth? The chest cavity?*

B. *Why is water, rather than air, used in this model?*

3. With your partner, plan how to build your model. Diagram your method in your notebook. When you have finished, show your proposal to your teacher, who will decide whether you are ready to begin constructing your model.

4. Using the syringe, balloon, and water, construct and operate your breathing model. Discuss the following questions with your partner:

A. *Does your model work as planned?*

B. *If not, what could you do to improve it?*

5. If necessary, try again to construct your breathing model.

6. After you have operated your breathing model, return your materials to the distribution center.

7. Now complete the table on Student Sheet 10.1.

Jazz trumpeter Dizzy Gillespie had great breath control. But it was his lungs and diaphragm, not those puffed-up cheeks, that really did the work!

REFLECTING ON WHAT YOU'VE DONE

1. Using Student Sheet 10.1, discuss the strengths and limitations of the syringe model of breathing.

2. In your science notebook, make a table similar in format to the one on Student Sheet 10.1. Title it "Using Models." Label the left column "Strengths" and the right column "Limitations." Reflecting on what you know about models in general, complete the table. Discuss your ideas with the class.

3. Using the information you read in the episode of "Spies" in this lesson, explain why it is better to breathe through your nose than through your mouth.

4. Take another look at the list you recorded in your science notebook at the start of the lesson. Revise it as necessary.

SPIES

THE SECOND JOURNEY BEGINS

Having recuperated from his trip through the digestive system, Bollo is ready for a new adventure. "Where are we going today?" he asks, looking at the human body map.

"We're going to explore the respiratory system—the system that humans use to breathe. Our map says that the departure point for this trip is just north of the mouth. It's the nose. Are you ready?"

"Let's go," Bollo replies.

The spies prepare to depart for a trip through the respiratory system. Starting point? The nose!

Pairs

"We already have a choice. Humans have just one mouth, but two nostrils. Do we go left or right?"

"It doesn't matter, because they both lead to the same place," says Peppi. "But now that you've mentioned it, this is something I want you to think about. In many cases, the human body is designed in pairs: two nostrils, two eyes, two ears, and so forth. Why do you think that is?"

"To make people look better?" says Bollo. "I've seen pictures of those one-eyed monsters."

"That's one explanation. Pairs *do* provide balance. Keep the idea of pairs in mind as our travels continue. Maybe you'll think of another reason."

Into the Nose

A whoosh of air draws Peppi and Bollo into the left nostril. They find themselves in a moist, dark, warm place.

"This is the nasal cavity," says Peppi. "It plays sort of the same role for the respiratory system that the mouth does for the digestive system. But here, we're dealing with air, not food."

Achoo!

Bollo and Peppi find themselves flying out of the nose.

"Wow! What happened?" says Bollo. "I nearly lost my cap."

"Joanne sneezed!" replies Peppi. "Air has tiny particles in it—dust, pollen, germs, and other things. One job of the human nose is to filter out those particles and to get rid of the ones that are irritating. When something really irritates the delicate lining of the nose, humans sneeze.

"Sneezes are caused by a sudden contraction of the muscles of respiration. When that happens, air bursts from the lungs. Sneezes can be powerful—I once clocked a sneeze at 100 miles per hour! Coughs are similar to sneezes, but they originate lower in the respiratory tract. They offer humans a second way of getting rid of those germs and dust."

Watch out, Bollo! Don't get trapped in the nasal hairs!

The Nose Knows

"The air looked fine to me," Bollo insists. "How does the nose detect dust?"

The two spies sail back up into the nostril for a second try. "Look around," says Peppi. "See those hairs waving back and forth? They catch the particles that are in the air that humans inhale.

"If the hairs don't catch everything, the body has a second line of defense. It's the mucous membrane, which is the lining of the nose. The membrane is covered by a thick substance called mucus. Feel it."

"Warm and slimy," says Bollo.

"That's right. Both these properties help the mucous membrane do its job. It traps particles that have sneaked by the nasal hairs. It also helps warm the air. Otherwise, the lungs would get a cold blast when humans are outside in winter."

"So the nose is the watchdog," says Bollo. "It's one of the body's lines of defense against dirt, germs, and other unwanted characters."

"Right," says Peppi. "But the nose has

another role, too. See that patch just over our heads? It's the olfactory membrane. The membrane is covered with cells that react to certain chemicals. When they meet up with a chemical, these cells send a message to the brain by way of the olfactory nerve. The brain decodes the message. And it becomes . . ."

"What?"

"A smell! All odors—from the fragrance of a fine perfume to the smell of a skunk—start when chemicals react with the receptor cells in the area of the nose. Smelling and breathing are related," says Peppi. "For example, when humans get a cold and their noses get stuffed up, they can temporarily lose their sense of smell."

In the Pipeline

Peppi and Bollo head downward toward a narrow tube. More hairs and mucus. ("More chances to get caught if the body thinks you might cause trouble," thinks Bollo.)

"Is that our old friend the esophagus below us?" asks Bollo.

"Yes. And remember the epiglottis? The 'safety valve' that keeps food and air going in the right direction? Look out ahead. It's important at this point, too."

"Yes, when we were investigating the digestive system, the epiglottis snapped shut, and we continued on our way to the stomach. This time, it's opening!" says Bollo.

The spies enter through the glottis, which is the opening to the windpipe.

"This is the larynx. It's also called the voice box," says Peppi. "The larynx is firm and stiff. Inside, stretching from the top to the bottom, are two pairs of thick bands."

"The vocal cords?" says Bollo, after sneaking a peek in Peppi's book.

"Absolutely right. There are two pairs. The first pair is the false vocal cords. They're not important, as far as speech is concerned. The pair below do the work. They are the true vocal cords.

"Human sound starts when air is pushed up from the lungs through the larynx. When the muscles of the larynx relax or contract, they make the vocal cords get longer or shorter. The more tension on the cords, the higher the pitch of the sound. As far as making specific sounds—saying 'cat' instead of 'that,' or 'dog' instead of 'log'— that's up to the mouth, lips, and tongue. They shape sound into words."

Bollo lands on one of the cords and jumps up and down a

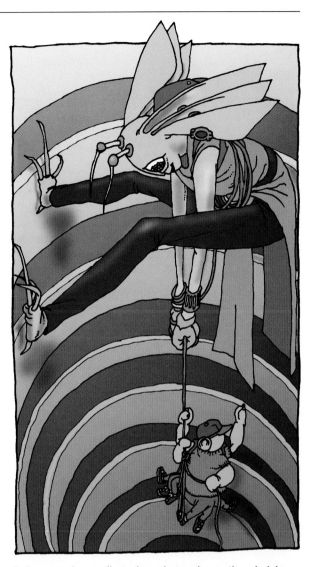

Bollo uses the cartilage rings that make up the windpipe as a make-shift ladder as he journeys down to the bronchi.

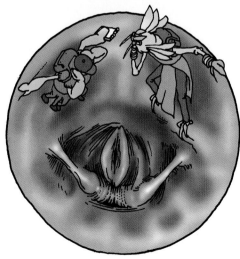

Look how the vocal cords open and close, lengthen and shorten, to help Joanne make different sounds!

few times. "Springy— sort of like a trampoline. I can see that they'd be able to stretch. But what makes sound loud or soft?"

"That depends on how much air passes through. If there is a lot of air, the sound is loud. Whispering needs just a tiny bit of air," replies Peppi.

A New Kind of Tree

"There's a lot more to the respiratory system than just moving air in and out. It has a role in the sense of smell and in speech. But let's get moving. I want to get to the center of the action— the lungs," says Bollo.

Peppi and Bollo continue down the windpipe, or trachea.

Peppi reviews the breathing process.

Just ahead, the road branches in two directions.

"We're approaching the lungs," says Peppi. "Each lung is served by a bronchus. (The plural of "bronchus," which is a Latin word, is "bronchi.") In humans, the bronchi are the beginning of something called the bronchial tree. Why do you think it got that name?"

"That's a no-brainer," says Bollo. "It is like an upside-down tree. The bronchus is like the trunk. It keeps branching out. Look at the tiny branches at the end!"

"Those are the bronchioles. The smallest ones are thinner than the finest hairs. At the end of each tiny branch are clusters of air sacs, or alveoli. It is through the membranes of these sacs that gas exchange takes place. And what gases are we talking about?"

"Oxygen and carbon dioxide," replies Bollo promptly.

"Right you are," Peppi says with a smile.

Time for Review

"Now before we go any farther, let's take time to review. Why do humans breathe in the first place?"

"Well, I know that humans breathe because they need oxygen," says Bollo. "And oxygen enters the body through the respiratory system. Humans also need to get rid of that carbon dioxide."

"Right. Now one of the remarkable things about this system is that it's all on 'automatic.' Nobody has to think about breathing, because it is controlled in a special location in the brain, called the medulla. When humans start to use more oxygen, for example, when they're working or playing hard and build up carbon dioxide, the muscle cells send a message to the respiratory center. 'Help! This carbon dioxide is killing us! We need more oxygen,' they cry. The brain tells the breathing muscles to speed up."

"So the lungs are muscles?" says Bollo.

"I can see how you might think so, but the lungs aren't that tough. The tough guys are the respiratory muscles, and there are two kinds.

"See those bands between the ribs? They're the intercostal, or rib, muscles. And that large, arched muscle directly below—the one looks like a sheet? That's the diaphragm.

"Those muscles work as a team. Watch and tell me what you see."

Bollo looks around. "The diaphragm is flattening out and the intercostal muscles are contracting and pulling the ribs upward and outward," he says.

"Right. That's what happens when humans inhale. The ribs go up and out, causing the lungs to expand. As the volume inside the lungs increases, air pressure inside drops, and new air rushes in. Now wait a minute."

"The diaphragm just moved up, and the intercostal muscles relaxed and fell in," says Bollo. "That

must mean that air is being exhaled."

"You've got the picture," says Peppi. "Humans normally breathe about between 10 and 14 times a minute. They breathe more often at times when they are using a lot of energy. The lungs hold about 6 liters of air.

"So now that we've taken a look at the big picture, let's concentrate on what happens when the air reaches the alveoli."

"Give me 15 minutes for a quick nap and I'll be ready for a close-up view of respiration," Bollo replies. □

How Much Air Can You Exhale?

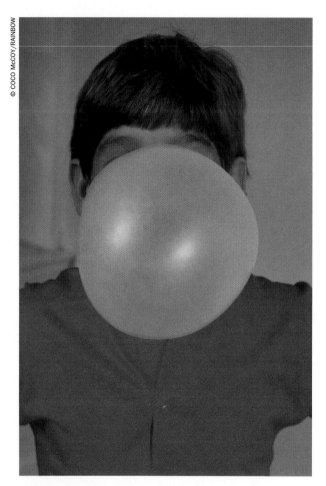

© COCO McCOY/RAINBOW

Bubble blowing is an enjoyable way to show your lung capacity!

INTRODUCTION

In the first part of this module, you discovered that the large surface area of the small intestine enables it to absorb large quantities of nutrients. The lungs also have a very efficient design. They contain millions of tiny air sacs, or alveoli. The alveoli make it possible for the lungs to hold and exchange a large quantity of air.

The amount of air you can exhale after taking a deep breath is a clue to how much air your lungs can hold. Just how much is that? You're about to find out!

OBJECTIVES FOR THIS LESSON

Use a sponge and water to model how the lungs hold air.

Graduate and assemble a device that measures the volume of exhaled air.

Determine how much air you can exhale after taking a deep breath.

Discuss the factors that affect how much air you can exhale.

Getting Started

1. You will work in pairs for this lesson. One student from each pair should collect the materials.

2. Remove a sponge, beaker, and graduated cylinder from the tote tray.

3. Pour 30 mL of tap water into the cylinder. Have your lab partner double-check the water level to make sure it is correct. Then pour the water into the beaker.

4. Using the forceps, pick up the sponge by an edge and dip it into the water in the beaker. Let the sponge absorb as much water as it can. Then, using the forceps, lift the sponge out of the water. Let the excess water drip into the beaker.

5. Continue to hold the sponge while your partner pours the water from the beaker into the graduated cylinder. Read the level of water in the graduated cylinder and subtract it from the amount of water you started with. Call this "The Amount of Water Absorbed by the Sponge." Record this amount in your science notebook.

6. Empty the water in the graduated cylinder into the sink or a large container.

MATERIALS FOR LESSON 11

For your group

- 1 tote tray
- 2 sponges
- 2 50-mL graduated cylinders
- 2 250-mL beakers
- 2 forceps
- 2 polyvinyl tubes
- 2 rulers, graduated in liters
- 2 plastic inserts
- 4 cardboard mouthpieces
- 2 black markers
- 2 strips of masking tape
- 2 large plastic storage bags
- 1 small resealable plastic bag
- 3 summary boxes
  Water (or access to sink)

7. Now hold the sponge over the beaker. Remove the sponge from the forceps. Squeeze as much water as you can from the sponge into the beaker. Then pour the water you have squeezed out of the sponge into the graduated cylinder. Read the level of water in the cylinder. Record this number as "The Amount of Water I Could Squeeze From the Sponge."

8. Subtract your answer in Step 7 from your answer in Step 5. Record this number as "The Amount of Water Remaining in the Sponge After Squeezing." Discuss the following questions with your group:

A. Was the sponge still wet after you squeezed out as much water as you could? Why or why not?

B. How do you think this activity relates to your lungs and the process of breathing?

9. Discuss your answers to the questions with the class.

Figure 11.1 *You're about to determine how much air you can exhale!*

Inquiry 11.1
Measuring How Much Air You Can Exhale

PROCEDURE

1. Follow along as your teacher reviews the Procedure. Then design a table in your science notebook to record your results.

2. Look at the ruler. Notice that it is marked in liters. Discuss the following questions:

A. How is this ruler different from other rulers you have seen?

B. What is one important difference between meters and liters?

C. How do you think the ruler was made?

3. Now it is time to prepare your apparatus. (When it is completed, it will look like the device shown in Figure 11.1.) The first step is to graduate the tubing, that is, to mark it with the appropriate units of measurement. To do so, follow these steps:

A. Unroll the tubing until approximately 50 cm are stretched out on your desk (see Figure 11.2). Note that one end is open and one end is sealed.

B. Place the ruler underneath the tubing so that the highest mark on the ruler is lined up with the sealed end of the tubing, as shown in Figure 11.2.

C. Label the seal at the closed end of the tubing "6.0 L."

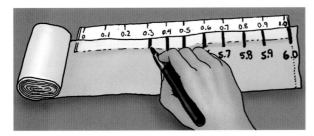

Figure 11.2 *Mark the tubing as shown here.*

D. With a marker, trace the lines that appear on the ruler onto the tubing. Label the lines in the same way in which the tube is marked. As you move down the tubing, repeat this procedure five more times. Your final mark should be "0.0 L."

E. Once you have reached the 0.0 L mark, measure another 0.1 L of tubing and place a mark at that spot. Cut the tubing at that point. The tubing will then be graduated to 6.0 L.

4. Begin to assemble your apparatus by putting a plastic insert into the tubing. Leave about 2.5 cm of it sticking out, as shown in Figure 11.3. Wrap the extra tubing securely around the mouthpiece and fasten the tubing securely to the mouthpiece using masking tape. Make sure that air cannot get between the mouthpiece and the tubing.

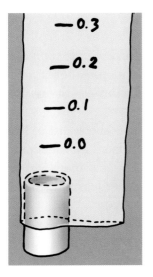

Figure 11.3 *Let about 2.5 cm of the plastic insert stick out of the tube.*

5. Now fasten the tubing to the plastic with masking tape (see Figure 11.4).

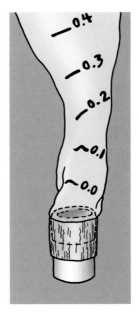

Figure 11.4 *Use masking tape to attach the plastic insert to the tube.*

6. Finally, place the cardboard mouthpiece snugly inside the plastic insert. When complete, your apparatus should look like the one shown in Figure 11.5.

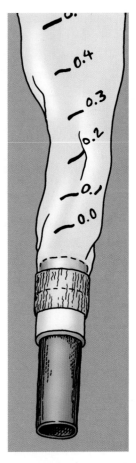

Figure 11.5 *Completed assembly*

7. Now perform the following steps to deter-mine how much air you can exhale.

A. Roll the tubing as close to the mouthpiece as possible.

B. While your partner holds the roll of tubing between both hands, take as deep a breath as you can. Slowly exhale as much air as you can into the tubing. Do not blow into the tubing too hard, because your partner will have trouble holding it as it unrolls. If you are the person holding the tube, move backward slowly as it unwinds.

C. Just before you take the tubing out of your mouth, put one hand on the plastic insert and the other around the tubing. Twist the tube at the end of the insert to keep air from escaping. Your partner will roll the excess tubing toward you.

D. Using the graduations on the tube, deter-mine how much air you exhaled. Record your answer in your science notebook.

E. Repeat the procedure once more. Record the information in your science notebook. Total the results of both trials and deter-mine the average.

F. Remove the cardboard mouthpiece and discard it. Replace it with a new one.

G. Reverse roles with your partner and repeat Steps A through E.

8. Remove the masking tape and plastic insert from the tubing. Put the plastic inserts in the small bag. Obtain four new cardboard mouthpieces from your teacher and put them in the same bag. Return the laminated ruler to the tote tray. Put the tape, tubing, and disposable cardboard mouthpieces into a trash can. Return the tote tray to the designated area.

REFLECTING ON WHAT YOU'VE DONE

1. In the "Getting Started" activity, you determined the following:

- the amount of water absorbed by the sponge
- the amount of water you could squeeze from the sponge
- the amount of water remaining in the sponge after squeezing

In your science notebook, answer the following questions:

A. Which of these amounts represents the air you could forcibly exhale after taking a deep breath?

B. Which of these amounts represents the air you could not forcibly exhale from your lungs?

C. Which of these amounts represents the total amount of air your lungs can hold when you take your deepest breath?

2. In your science notebook, list at least five things that affect how much air your lungs can hold. Discuss them with the class. Add any new ideas that your classmates suggest.

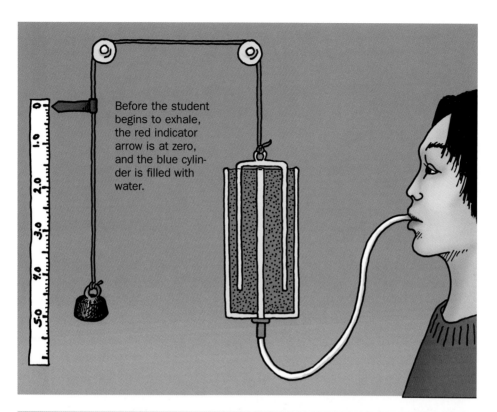

Before the student begins to exhale, the red indicator arrow is at zero, and the blue cylinder is filled with water.

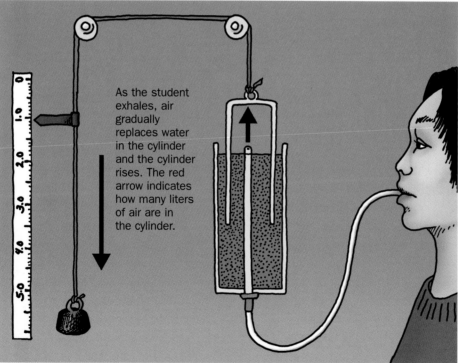

As the student exhales, air gradually replaces water in the cylinder and the cylinder rises. The red arrow indicates how many liters of air are in the cylinder.

The spirometer is used to measure vital capacity. The spirometer shown here is an early model. Modern spirometers operate electronically.

UP, UP, AND AWAY!

Ice and cold are not the only obstacles these mountain climbers face. Can you name another challenge of climbing?

You've probably heard of motion sickness and seasickness. But mountain sickness? What's that?

Mountain sickness is a condition that develops when the blood is not receiving enough oxygen. Mountain climbers may get mountain sickness, because the air at higher elevations contains less oxygen than the air at lower elevations.

The higher you climb, the thinner the air. If you were standing on the peak of a mountain that was 6000 meters high, the air pressure would be just half what it is at sea level. If your breathing did not change, you would be inhaling only half your normal supply of oxygen. Mountaineers who reach the peak of Mt. Everest, which is about 10,000 meters high, are breathing air that contains only about one-third the amount of oxygen present at sea level.

At high altitudes, the lungs can't deliver all the oxygen the body needs. As a result, the heart doesn't have enough oxygen-rich blood to distribute to the cells of the body.

The effects of an oxygen shortage are felt throughout the body. The symptoms include severe headache, fatigue, nausea, and shortness of breath. When experienced mountain climbers develop these symptoms, they know it's time to return to a lower level or to get out their oxygen tanks.

How can you prevent mountain sickness? By proper conditioning. Your body will adjust to higher altitudes if you give it enough time. Shortly after ascending to a higher altitude, you begin to breathe more deeply. Your respiratory system works harder. All the alveoli fill up with each breath. Your heart rate increases, too.

This doesn't happen at sea level, where the lungs don't have to work so hard. Experienced mountaineers take it easy. For example, they may climb about 300 meters higher each day. They usually return to a lower altitude to sleep. This is because people breathe more slowly at night than during the day. This means that they get less oxygen at night, even under the best of circumstances.

Even if you're well conditioned, your body cannot perform as well at high altitudes as it can at sea level. What's more, you don't actually need to get mountain sickness to experience some of the effects of high altitude. Athletes, for example, are very aware of altitude. An Olympic athlete who has trained at sea level will be at a definite disadvantage if the competition takes place 1500 meters higher! □

Mountain climbers carry a supply of oxygen and a mask.

Dr. Heimlich's Lifesaving Maneuver

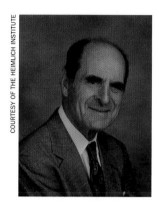

COURTESY OF THE HEIMLICH INSTITUTE

Dr. Henry Heimlich

It's not often that a doctor's name becomes a household word while he's still alive. But this is the case with chest surgeon Henry J. Heimlich. Dr. Heimlich is the inventor of the Heimlich maneuver. He wrote about it in the 1970s, and by 1980, the term "Heimlich maneuver" had entered the dictionary.

The Heimlich maneuver is used to dislodge an object from the windpipe of a person who is choking. If someone cannot breathe or is not getting enough air into his or her lungs, quick action is essential. Without oxygen, the brain will suffer irreversible damage within 4 to 6 minutes.

Before Dr. Heimlich came up with this idea, doctors had been recommending a very different technique to remove objects from the throat. Their suggestion was to reach into the person's throat or slap the person on the face to force the object out. These strategies often had the wrong effect: They lodged the object deeper in the throat instead of forcing it up.

Dr. Heimlich's solution was to apply a force from below. This would push the object out, he reasoned. The technique he recommended uses the lungs as a bellows. The air in the lungs moves quickly up into the windpipe and exerts a force on the object.

Dr. Heimlich's maneuver has saved the lives of more than 100,000 people, including former President Ronald Reagan and actress Elizabeth Taylor. ☐

© D. YOUNG-WOLFF/PHOTOEDIT./PNI

The Heimlich maneuver has saved many lives. But don't try it unless you have proper training.

Recipe for Energy— Cellular Respiration

Many athletes "carbo load" before a game or meet. Is this a good idea? Why?

INTRODUCTION

You might think of nutrients and oxygen as two of the ingredients, or raw materials, that are needed to maintain life and growth. But how do the cells use these two substances? What happens to the wastes produced as a result? These are the questions you are about to explore.

In this lesson, you will observe what happens when a fuel combines with oxygen. The fuel you

OBJECTIVES FOR THIS LESSON

Use a model to determine the raw materials and products of combustion.

Recognize that combustion is a form of oxidation.

Recognize that cellular respiration is a form of oxidation.

Compare and contrast combustion and cellular respiration.

Use an indicator to detect a gaseous waste common to combustion and cellular respiration.

Determine a form of energy released during exhalation.

Observe and document the flow of inhaled and exhaled air through a model.

Use the model to discover whether a gaseous waste product of cellular respiration is present in inhaled and exhaled air.

Determine whether the gaseous waste product of cellular respiration can pass through a membrane.

will use is candle wax, and the burning process is called combustion, a form of oxidation during which a great deal of energy is released very rapidly. You will identify two of the waste products of combustion. You will then determine whether either of these products is similar to those produced during another form of oxidation called cellular respiration by exploring whether they are present in the air you inhale and exhale. You'll also determine which ingredient is common to combustion and cellular respiration. You'll combine these observations to develop a brand-new recipe—a recipe for cellular respiration!

OXIDATION—ONE PROCESS, TWO FORMS

By now you know that oxygen is a gas—a gas that humans inhale with every breath they take. In this lesson, you will be studying what happens when a substance combines with oxygen. This process is called oxidation. You will examine two very different forms of oxidation. The first is called combustion. Combustion is a form of oxidation that is accompanied by the rapid release of energy, in the form of heat and light. This rapid burning produces waste products.

Cellular respiration is another form of oxidation; in other words, it is also a process by which a substance combines with oxygen. Cellular respiration, like combustion, also produces waste products. Cellular respiration is not nearly as dramatic as combustion. Nevertheless, it is a process that is essential to life—one that goes on continually in just about every cell of the body.

MATERIALS FOR LESSON 12

For you
1 copy of Student Sheet 12.2: Inhaled Versus Exhaled Air
1 copy of Student Sheet 12.3: Venn Diagram: Combustion and Cellular Respiration
1 pair of safety goggles

For your group
1 tote tray
1 test tube rack
1 candle
1 wooden block
1 piece of aluminum foil
1 Inhaled Versus Exhaled Air Apparatus
2 large test tubes
2 medium test tubes
2 pairs of scissors
2 plastic funnels
4 drinking straws
1 dispensing bottle of bromthymol blue solution
2 50-mL graduated cylinders
2 membranes
2 pieces of string
2 thermometers
1 250-mL beaker, unlabeled
1 250-mL beaker labeled "Carbonated Water"
1 250-mL beaker labeled "Tap Water"
2 red pencils
2 blue pencils
Water (or access to a sink)

Getting Started

1. Have one person from your group pick up your materials. You will work as a group for the "Getting Started" activity.

2. Review the Safety Tips with your teacher.

SAFETY TIPS

Do not get near the candle flame. If you have long hair, tie it back.

Always wear safety goggles when you are working with bromthymol blue.

For sanitary reasons, do not share the straws.

If the bromthymol blue solution comes into

contact with your eyes or skin, rinse it off immediately with water.

To prevent the liquid from spilling, inhale and exhale slowly and gently through the straw mouthpiece.

Notify your teacher immediately of any chemical spills.

3. Using the graduated cylinder, measure 15 mL of bromthymol blue solution. Pour the solution into the unlabeled 250-mL beaker. Put the wooden block in the beaker and place the candle on top of the block (see Figure 12.1). Set it in a place where everyone in your group can see it.

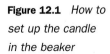
Figure 12.1 *How to set up the candle in the beaker*

4. After your teacher lights the candle, watch what happens as it burns. Answer the following questions in your science notebook:

A. What is the candle made of?

B. In addition to the bromthymol blue, the candle, and the block of wood, what other substance is present in the beaker?

C. What two ingredients, therefore, are involved in the burning of the candle?

D. What is released very quickly as the candle burns?

E. If you wanted to put out the candle, what gas would need to be removed from the candle's environment?

5. Now set a square piece of foil over the top of the beaker. Fold the edges of the foil down over the sides of the beaker.

A. What happened to the flame when you covered the beaker with foil? Why?

B. What gas do you think remains in the covered beaker when the flame has gone out?

C. What else do you see in the beaker?

6. Do not remove the foil. Swirl the solution gently around the bottom of the beaker for 1 minute, as shown in Figure 12.2.

Figure 12.2 *Swirl the solution gently in a slow, circular motion.*

A. *What happened to the bromthymol blue solution?*

B. *What substance do you think caused this change?*

C. *Bromthymol blue is a special chemical called an indicator. On the basis of what you discovered during this inquiry, how would you define an indicator? What other indicators have you used in previous lessons in this module?*

7. Follow your teacher's instructions for cleanup.

8. Review this activity with your teacher. Agree on the two ingredients necessary for combustion of the candle and the products that were released during the process of combustion.

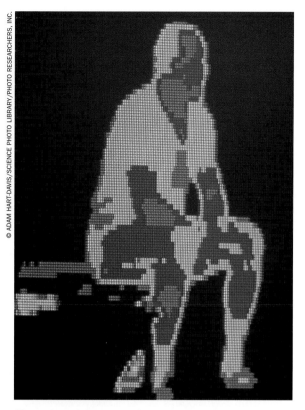

These heat images, or thermograms, show a man immediately before (above) and after playing a game of handball. What do you think the color changes indicate?

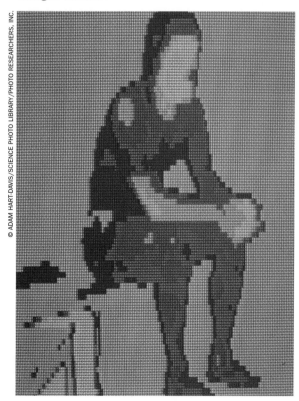

Inquiry 12.1
Investigating Cellular Respiration

PROCEDURE

1. Now it is time to apply what you observed during "Getting Started" to an important process that goes on in the human body. How is combustion similar to another form of oxidation—a process called cellular respiration—that occurs in your cells? How are the two processes different? Before you begin to explore these issues, answer the following questions in your science notebook:

   A. *What did you see as the candle burned that would be absent in the type of oxidation that occurs in your body cells?*

   B. *What ingredient of combustion do you think is also an ingredient for cellular respiration?*

   C. *On the basis of what you discovered in the first section of this module, what do you think is the second essential ingredient for cellular respiration?*

2. When a candle burns, you can easily observe the energy that is released in the form of heat and light. But what evidence can you find to show that your body releases energy during cellular respiration? Working with your partner, follow these steps to find out.

   A. Use a graduated cylinder to measure 5 mL of the tap water that your teacher has provided for you. Pour the water into a test tube. Place the thermometer into the water and allow it to sit for about 30 seconds. Read the temperature. Record this as the "Starting Temperature." Record it in your science notebook.

   B. Leave the thermometer in the water. Put a straw in the test tube, bend the straw slightly toward you, and blow steadily through it into the water for about 2 minutes. Read and record the temperature of the water.

3. Discuss the following questions with your partner. Then answer them in your science notebook.

   A. *Did the temperature of the water change after you exhaled into it? If so, what was the change?*

   B. *What is temperature a measurement of?*

   C. *What would a change in the temperature of the water indicate about your exhaled air?*

   D. *How would you relate this change to a product of cellular respiration?*

4. Follow your teacher's instructions for cleanup.

Inquiry 12.2
Using a Model to Show Evidence of a Waste Product of Cellular Respiration

PROCEDURE

1. In Inquiry 12.1, you observed evidence of a product of cellular respiration in your exhaled air. In this inquiry, you will use a model to identify a waste product of cellular respiration. Watch as your teacher displays a transparency showing the model.

2. Work with your group to explore how the model works and complete the inquiry by following these steps:

A. Remove the stoppers from the two test tubes and fill each tube with 10 mL of water from the tap. Replace the stoppers on the test tubes.

B. Place the tubes in the test tube rack.

C. Create a mouthpiece by cutting a straw in half and inserting one of the pieces of the straw into the end of the rubber tube extending from the plastic Y-tube (see Figure 12.3).

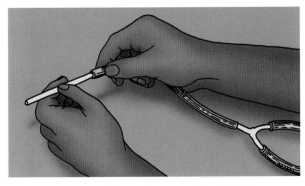

Figure 12.3 *How to insert the straw into the tube*

D. Have one student in your group breathe through the straw mouthpiece until everyone can observe the path of inhaled and exhaled air through the model. The student who is breathing through the apparatus should try to inhale and exhale at the same rate and with the same amount of force.

E. Color code the path that inhaled and exhaled air take through the apparatus on the illustration on Student Sheet 12.2. Use a red pencil for inhaled air and a blue pencil for exhaled air. Answer the following question in your science notebook:

On the basis of your observations, what is the main thing about the way the apparatus is assembled that seems to determine whether inhaled or exhaled air bubbled through the water in the test tube?

3. Remove the stoppers and empty the water into the sink.

4. Using the graduated cylinder, measure 10 mL of bromthymol blue solution and pour it into one of the large test tubes. Then do the same thing with the second test tube. Replace the rubber stoppers on the two test tubes.

5. Using one-half of a straw as a mouthpiece, have a different group member repeat the exercise until you see a definite color change in one of the test tubes.

6. Follow your teacher's directions for cleaning up and returning materials.

7. Answer the following questions in your science notebook:

A. Did you see a color change in the bromthymol blue solution in either test tube after someone in your group had breathed into it?

B. In which test tube did the color change occur?

C. What substance caused this change?

D. During what process in your cells is this substance released?

8. You have now determined two of the products of cellular respiration that can be found in your exhaled air. Discuss these products with your group. Agree on what they are.

Inquiry 12.3
Exploring the Movement of Carbon Dioxide Through a Membrane

PROCEDURE

1. In Inquiry 12.2, you found evidence of carbon dioxide in your exhaled air. You found no evidence of carbon dioxide in your inhaled air.

   How do you think the carbon dioxide was carried to your lungs?

2. Your blood transports oxygen to body cells, where it is exchanged for carbon dioxide. But how do both gases get past the membranes of your body cells? Work with your partner and follow these steps to find out.

   A. Place a large test tube in the rack.

   B. Pick up a wet membrane from the beaker. Tie one end of the membrane with string. (Refer to Step 3B of the Procedure in Lesson 6 if you need directions on how to do this correctly.)

   C. Measure 13 mL of carbonated water in the graduated cylinder. Use a funnel to pour it from the cylinder into the membrane (see Step 3D of the Procedure in Lesson 6 for additional directions).

   D. Rinse the outside of the membrane with water. Hold the membrane by the open end and lower it into a large test tube. Place the open end of the membrane down the side of the test tube. Place the test tube in the test tube rack.

   Why do you need to rinse the outside of the membrane with water?

E. Measure 7 mL of bromthymol blue solution in a graduated cylinder. Pour it into the test tube containing the membrane. Let the test tube and membrane sit for about 30 seconds.

F. Observe the contents of the test tube. Use the following questions to guide your observations. Record your answers in your science notebook.

1. Did the color of the bromthymol blue solution change? If so, what caused the color change?

2. What does this color change tell you about the ability of carbon dioxide to pass through a membrane?

3. Through what process does carbon dioxide pass through membranes?

4. How do you think oxygen passes through membranes?

3. Return your materials to the designated area. Put the membranes back in the beaker of water. Make sure they are submerged.

4. Work with your group to complete the Venn diagram on Student Sheet 12.3.

REFLECTING ON WHAT YOU'VE DONE

1. With the class, discuss the responses you have written on Student Sheets 12.2 and 12.3.

2. On the basis of what you have learned in this lesson, answer the following questions in your science notebook.

A. What are the ingredients, or raw materials, for cellular respiration?

B. What are the products of cellular respiration?

C. Which product of cellular respiration enables humans to perform life activities?

D. Which of the products of cellular respiration are wastes? How are they transported to the lungs?

E. Inhaled air contains about 0.03 percent carbon dioxide and 20 percent oxygen. Exhaled air contains about 4 percent carbon dioxide and 16 percent oxygen. On the basis on what you've learned in this lesson, explain why this is so.

F. Summarize in one sentence the "recipe" for cellular respiration. Mention both the raw materials and the products.

G. Why do you think people need to eliminate the wastes of cellular respiration from the body?

3. Discuss your answers to these questions, as well as the other observations you have made during this lesson, with the class.

POLIO:
Machines and Medicine Control a Killer

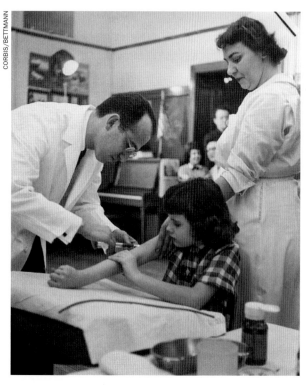

Dr. Jonas Salk administers his polio vaccine to an 8-year-old girl during the field trials of the new vaccine in 1954.

Dr. Albert Sabin in his laboratory at the University of Cincinnati College of Medicine in Ohio

You're lucky. Before you started school, your doctor gave you a medicine (called a vaccine) to protect you from polio. Until the middle of the 20th century, the polio vaccine didn't exist, and each year thousands of people got sick. In 1950, more than 33,000 Americans got polio, which is also called infantile paralysis. Most of them were children. Some of them became crippled or died from the disease.

In 1953, Dr. Jonas Salk invented a vaccine that protects people from polio. The vaccine is made from the same virus that causes polio, but the virus in the vaccine has been killed. It won't make you sick. It tells your body's immune system what the polio virus "looks like." Then, if live polio virus should get into your body, special cells in your blood system would recognize it as dangerous and rush to destroy it.

The polio vaccine that Dr. Salk developed is injected with a needle. You probably took your polio vaccine by drinking it. This is called an oral vaccine. The oral polio vaccine was developed by Dr. Albert Sabin.

Many people with polio are not extremely sick. They just feel like they have the flu. But sometimes the polio virus travels to the nerves in the spinal cord, which relays messages to the muscles. The nerves cannot send signals to control the muscles. If the virus damages the nerves going to the legs, for example, the patient's legs become weak. People with this kind of polio need crutches or wheelchairs to get around.

Sometimes polio damages the nerves that control breathing. These patients need help getting air in and out of their lungs. In 1926, an

American engineer named Philip Drinker invented a machine to help polio victims breathe. The machine, called an iron lung, is a metal box big enough to hold a polio patient. The patient's head sticks out through a rubber collar that keeps air from getting in or out. At the other end is a flexible diaphragm that is moved back and forth by a motor. As the diaphragm is pulled outward, the pressure inside the iron lung decreases. A force is created that expands the patient's chest. This draws air through the patient's mouth and into the lungs. When the diaphragm is pushed inward, the pressure inside the iron lung increases, the chest is pushed in, and air is forced out of the patient's lungs.

The iron lungs proved that many people could survive polio if they got help breathing. Since the iron lung was invented, doctors and engineers have worked together to make iron lungs that are smaller and easier to use. Today, there are mechanical ventilators (machines that help people breathe) that are small enough to be carried around like a suitcase. These ventilators gently push air into a patient's lungs through small tubes.

Thanks to the pioneering work in vaccine development of two doctors—Salk and Sabin—polio has virtually disappeared in many countries, including the United States. And thanks to the work of scientists like Phillip Drinker and others, mechanical ventilators are saving the lives of many people who need breathing support—from tiny premature infants to older persons with lung disease. ❑

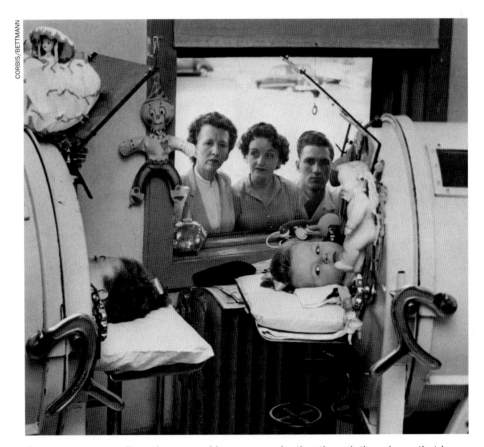

CORBIS/BETTMANN

These two young polio patients are able to see each other through the mirrors that have been mounted over their heads. In the middle part of the 20th century, thousands of children developed polio every year. This photograph was taken in Tennessee in 1952.

SPIES

WHY SO MANY?

Peppi and Bollo are ready to continue their exploration of the human respiratory system.

"Did you say that there are *hundreds of millions* of alveoli?" asks Bollo.

"Right. Most humans have about 300 million of them. If you laid the lung tissue of an adult human out flat, it would cover the area of a tennis court."

"Why are there so many?"

"Remember when we were inside the small intestine and saw the villi?" asks Peppi. "What did you find out about them?"

"We learned that the large surface area makes absorption work more efficiently," says Bollo. "Hey, I've got it—I bet the alveoli help make gas exchange easier."

"Right again! The alveoli make gas exchange more efficient because they provide so much space for it to happen. They help the body get oxygen fast. Let's hop aboard an oxygen molecule and see just how fast."

Peppi and Bollo are pulled farther and farther down the bronchial tree. Ahead, they see an alveolus. It's covered by tiny blood vessels. ("Capillaries," says Peppi.) From inside, the vessels look like a net. Some of the capillaries are dark red; others are bright red.

Still traveling aboard an oxygen molecule, the spies approach the wall of the alveolus. Suddenly, they burst through the wall of the alveolus, and then through the wall of a blood vessel. (It doesn't take much effort, because the wall of each organ is one cell thick.) They find themselves attached to a strong new partner—a red blood cell.

"We made it," says Peppi. "Welcome to the bloodstream. The oxygen is on its way to the cells. And because the cells can't live without oxygen, I'm sure

All aboard for a ride on a red blood cell!

we'll be welcome guests. But first, we've got to take a side trip through the heart. Our route is a pulmonary vein."

Peppi and Bollo are moved into the heart and are pumped out into a huge artery.

"Where are we going to end up?" asks Bollo.

"It doesn't matter," says Peppi. "The important thing is that all cells need oxygen to survive. When the oxygen arrives in body cells, tiny structures in the cells, called mitochondria, take over. It is in these tiny, sausage-shaped structures that the real work takes place. The oxygen combines with nutrients (usually glucose) from digested food in a form of oxidation called cellular respiration. There's no flame, of course, but this process releases the energy that humans need to live.

"Carbon dioxide and water are produced during cellular respiration, too," Peppi continues. "They're considered waste products. They pass into the blood. The carbon dioxide is carried back to the lungs and released during exhalation."

On the Way Home

Once they've made their stop at a body cell and unloaded the oxygen, Bollo notices that things have changed. That bright-red blood is now dark red.

"That color change is important. Bright-red blood has oxygen in it. Dark-red blood has less oxygen and more carbon dioxide," explains Peppi.

The spies, floating in the plasma with some of the carbon dioxide, move back toward the

Peppi points out the mitochondria in a cell.

heart, then move through the pulmonary artery and toward the lungs.

Again, they move through two thin walls, but this time the process is reversed. They're moving back into the lungs from the bloodstream.

The diaphragm relaxes, the rib muscles let the chest cavity deflate, and whoosh! Up and out! Peppi and Bollo are back in civilization.

Thinking It Over

"Breathing is more complicated than I thought," says Bollo. "It's not just a matter of in and out."

"Breathing is a physical process, and it's something that we can see people do. Sometimes we can even hear it! But breathing also is a necessary step leading to a chemical process called cellular respiration. Cellular respiration is the process by which glucose is broken down in cells in the presence of oxygen to supply energy for life activities."

"Is glucose the only nutrient used for energy?" asks Bollo.

"Some cells depend almost entirely on glucose," replies Peppi. "Others seem to prefer fats as a source of energy. Sometimes, when there are not enough amino acids in a cell to build a complete protein, the protein, too, can be used for energy or converted to fat."

"So the human body really makes use of those nutrients!" says Bollo.

"It sure does! And, speaking of food, I'm hungry. And my backpack is empty. Do you have any leftovers?" ☐

13

Releasing Energy From Food

Which has more calories—a toasted marshmallow or an untoasted one? By the end of this lesson, you should be able to answer this question.

INTRODUCTION

In Lesson 12, you explored how oxygen and nutrients react chemically in your cells to release energy. But do all foods release the same amount of energy? Do you get the same amount of energy from a 4-ounce hamburger patty as you do from 4 ounces of spaghetti or of chocolate fudge ice cream? How is the energy in food measured?

In this lesson, you will explore and compare the energy value of two different foods. You'll read about how scientists measure the energy in food and about the units in which the energy in food is measured.

By the time you are finished, "counting calories" may have a new meaning!

OBJECTIVES FOR THIS LESSON

Compare the energy value of marshmallows and walnuts.

Read about and discuss how the energy in food is measured.

Construct a definition for the word "calorie."

Getting Started

1. With your teacher, develop a list of things that you already know about calories.

2. Read "Counting Calories: Bombs Away!" and "Large or Small?" (on page 112) Discuss these reading selections with the class. On the basis of your reading and discussion, revise your list.

COUNTING CALORIES: BOMBS AWAY!

You may have heard that food contains calories. You can't see them or taste them. What are calories, and how do scientists measure them?

A calorie is not a "thing." It is a unit of measure, just like a kilogram or a meter. Kilograms measure mass; meters measure length. Calories measure heat energy. One calorie is the amount of heat energy needed to raise the temperature of 1 gram of water 1 °C.

To find out how much energy is in food, scientists measure how much heat the food produces when it is burned.

Most foods won't burn if you try to light them with a match. But they will burn in a bomb calorimeter! A bomb calorimeter is a device that scientists use to measure calories in food.

First, the food is ground up and dried. A carefully weighed sample is placed inside a strong steel container about the size of a coffee mug (the "bomb"). The container is sealed. Oxygen is pumped in until the pressure is extremely high.

(continued)

For you
1 pair of safety goggles

For your group
1 tote tray
4 small marshmallows
4 walnut pieces
1 candle
2 dissecting needles
2 thermometers
2 large test tubes
1 test tube rack
2 test tube clamps
2 50-mL graduated cylinders
1 400-mL beaker of water
1 250-mL beaker of water
4 paper towels
1 pie pan
Water (or access to a sink)

Counting Calories (continued)

The bomb is placed in an insulated container of water that has been carefully measured.

An electric current is sent through wires into the bomb. This creates a spark that ignites the food sample and oxygen. The explosive reaction burns all the food and heats the bomb. The bomb heats the water. A thermometer tells the scientist how much warmer the water has become as a result of the heat released from the food.

The scientist knows exactly how much water was in the container and how much food was inside the bomb. The scientist also knows that it takes 1 calorie to raise the heat of 1 gram of water 1 °C. With this information, it's easy to count the calories.

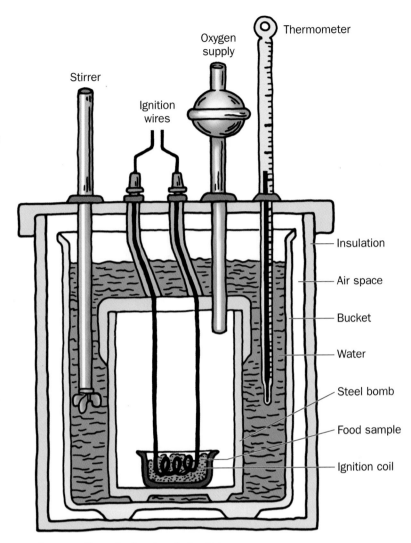

Cross-section of a bomb calorimeter.

Large or Small?
Calories come in two sizes: large and small. A small calorie is the amount of energy that will raise the temperature of 1 gram of water 1 °C. For the sake of convenience, the energy value of food is measured in large calories, or kilocalories. Each large calorie is equal to 1000 small calories.

That's enough energy to heat 1 kilogram (1000 grams) of water 1 °C.

When scientists talk about the energy in food, they spell Calorie with a capital "C" to make it clear that they mean kilocalories, not small calories. The calories provided on the Nutrition Facts labels on packaged foods and in calorie lists are kilocalories.

Inquiry 13.1
Comparing the Energy Released by Marshmallows and Walnuts

PROCEDURE

1. Have one member from your group pick up a tote tray. You will work in pairs for this activity.

2. After you have read the Procedure for this inquiry with the class, design a data table in your science notebook on which to record your results. After you do, record your answers to the following two questions in your notebook:

   A. Which food do you think will raise the water temperature more—marshmallows or walnuts?

   B. Why do you think one food might raise the temperature more than the other?

3. With your teacher, review the Safety Tips for this activity.

4. Follow these directions to determine the relative energy value of a marshmallow.

A. Pour exactly 20 mL of water from the 400-mL beaker into the graduated cylinder. Then pour the 20 mL of water from the cylinder into the test tube. Put the test tube clamp securely around the test tube.

SAFETY TIPS

Wear safety goggles while performing this inquiry.

Consider the foods you will use in this lesson as laboratory chemicals. Do not eat them or put them in your mouth.

Keep the end of the needle with the burning food facing upward.

Keep the test tube pointed away from you and classmates while you are heating the water.

Place the candle on the pie pan so that it will catch bits of burning food that may fall off the needle.

When you have finished burning a marshmallow or walnut, dip it into the small beaker of water to cool it. Then remove the food from the needle with a paper towel.

Be careful with the candle. Let it cool for a while before you pick it up again.

B. Insert the thermometer into the test tube. Allow it to rest in the water for 30 seconds. Read the temperature and record it on your data table (see Figure 13.1).

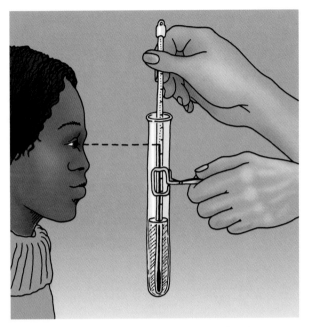

Figure 13.1 *Read the thermometer at eye level.*

C. Place a candle in a pie pan. Your teacher will light a candle for your group.

D. Stick the needle into a marshmallow and hold it over the flame until the marshmallow ignites (see Figure 13.2). As soon as the marshmallow ignites, have your partner extinguish the candle flame and move the candle to the side of the pie pan.

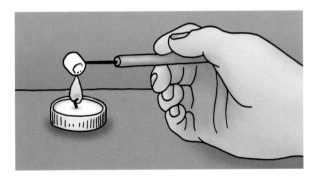

Figure 13.2 *Toasted marshmallow!*

E. Immediately place the marshmallow so that the tip of the flame is touching the bottom of the test tube. Hold it there until the flame goes out (see Figure 13.3).

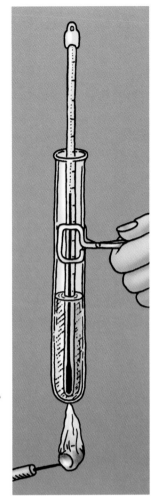

Figure 13.3 *Hold the marshmallow under the test tube.*

F. Wait 25 seconds. Read the temperature of the thermometer in the test tube. Record the temperature on your data table.

G. Dip the food in the 250-mL beaker of water for a moment to allow it to cool. Remove the remains of the marshmallow from the needle with a paper towel.

H. Rinse the test tube and refill it with 20 mL of fresh water from the 400-mL beaker. Repeat Steps B through G. Record your data. Calculate the average temperature rise of the water.

5. Now determine the relative energy value of a piece of walnut.

A. Stick the dissecting needle into a walnut piece by gently twisting the wooden handle of the needle back and forth with your thumb and forefinger until the point of the needle is securely fixed in the walnut.

B. As shown in Figure 13.4, follow the same procedure that you carried out with the marshmallow. Record your data in your science notebook.

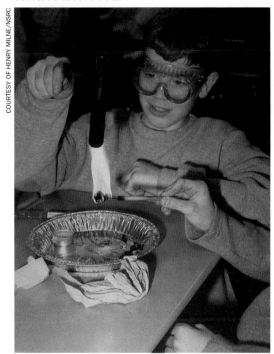

COURTESY OF HENRY MILNE/NSRC

Figure 13.4 *The pie pan will catch any residue that falls from the burning walnut.*

6. Follow your teacher's directions for cleanup.

SAFETY TIP

The candle will remain hot for a while after the flame goes out. Be especially careful with the melted wax. Keep the candle on the pie tin when you return it to the tote tray.

REFLECTING ON WHAT YOU'VE DONE

1. On the basis of what you discovered in this inquiry, answer the following questions in your science notebook:

A. Which of the two foods has a higher energy content? What evidence do you have for this?

B. What do you think the process you observed during this inquiry might have to do with calories?

C. How is the process you observed during the inquiry similar to cellular respiration?

D. How is the process you observed in this lesson different from the process of cellular respiration? (Refer to the Venn diagram that you drew in Lesson 12 if necessary.)

2. With your teacher, take another look at the list your class developed at the beginning of this lesson. Revise the list as necessary.

GO FOR THE BURN

Exercise is part of a healthy lifestyle.

"Go for the burn!" Did you ever hear this expression?

Athletes and fitness experts use it to describe the fiery sensation that develops in your muscles when you use them intensely. You might feel such a "burn" when you've just climbed a long flight of stairs or pedaled your bicycle up a hill.

Every day, however, your body is going for a different kind of "burn." The burn is a process that goes on constantly, whether you're exercising or sleeping. It involves the oxidation (burning) of foods, which is measured in calories. A calorie is a unit of heat energy. Nutritionists measure how much energy we get from food in terms of calories.

People who want to stay healthy know that it's important to control their weight. That means they have to know about calories. If they want to lose weight, they have to burn more calories than they consume. This can be done through a combination of exercise and a balanced diet.

Teenagers need more calories than children or older adults do. If you're a boy between 11 and 14 years of age, you need about 2500 calories a day; a girl of this age needs about 2200.

These numbers are averages. You may need more calories or fewer, depending on your size and your daily activities. For example, if you play soccer, you may need as many as 4000 calories a day. Some of these calories are used to fuel your athletic activity; the rest of them give your body the energy it needs to function and grow. A soccer player may burn as many as 400 calories an hour. A swimmer uses about 300 calories an hour.

But if you're a "couch potato," beware! Sleeping takes only 60 calories an hour. Reading takes about 100.

Counting calories is one way of making sure your body has enough fuel to burn and to keep you healthy. It is important to learn about the caloric content of foods. Lists of calories are available from many sources. Your local grocery store may have them. You can read the Nutrition Facts labels on canned or packaged foods. The Internet has many sources of information on this topic.

Good nutrition is much more than just counting calories. If your goal is to consume 2200 calories a day, you can't just eat 10 220-calorie candy bars! You have to pay attention to the types of food you eat.

The calories in foods such as candy bars come mainly from fat and sugar. These foods lack the fiber, vitamins, and minerals your body needs. Nutrition experts recommend that teenagers limit their fat consumption to less than 30 percent of their daily food intake. This would amount to about 83 grams of fat.

If you want to know how many grams of fat are in your favorite food, read the Nutrition Facts label. For example, one ounce of potato chips (about 20 chips) has about 150 calories and 9 grams of fat.

It doesn't take many calories to change a TV channel!

COURTESY OF THE USDA CENTER FOR NUTRITION POLICY AND PROMOTION

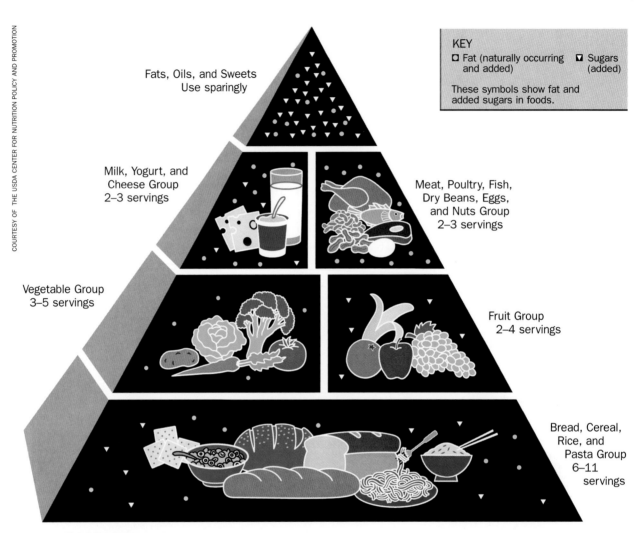

KEY
☐ Fat (naturally occurring and added)   ☑ Sugars (added)

These symbols show fat and added sugars in foods.

Fats, Oils, and Sweets
Use sparingly

Milk, Yogurt, and
Cheese Group
2–3 servings

Meat, Poultry, Fish,
Dry Beans, Eggs,
and Nuts Group
2–3 servings

Vegetable Group
3–5 servings

Fruit Group
2–4 servings

Bread, Cereal,
Rice, and
Pasta Group
6–11
servings

The Food Guide Pyramid

Cereal or potato chips? Maybe your mother does know best.

A cup of oat cereal has the same amount of calories, but only 2 grams of fat.

If you're concerned about a healthy diet, limit your fat intake and focus on eating starches and proteins. The Food Guide Pyramid (page 118), developed by the U.S. Department of Agriculture, can help you make daily food choices. It shows how many servings of each of the basic food groups you should eat in a day.

Carbohydrates are found in fruits, many vegetables, and flour or cereal products. About 55 percent of your diet should come from carbohydrates. They provide energy that helps your body use proteins. Proteins, which are found mainly in meat, poultry, eggs, and dairy products, help build muscle and bone tissue. Protein should make up about 15 percent of your food intake.

Whether you're a prize-winning athlete, a couch potato, or somewhere in between, it's important to know how many calories you need and what foods to eat for a well-balanced, nutritious diet. Once you do this, your weight can be within your control. ❑

14
The Pumping Heart

In many cultures, the heart is a symbol of love. For this reason, it is a popular organ around February 14! But is the heart really shaped like a valentine?

INTRODUCTION

You now know the ingredients that cells need to release energy—oxygen and nutrients. How do these essential ingredients get to the cells? How are the waste products—carbon dioxide and water—carried away?

Your circulatory system does the job. What powers your circulation? It's a special type of double-action pump—known as the heart.

In this lesson, you will use a model to explore how the heart works. You will also identify the most important structures of this organ. As usual, Peppi and Bollo will be on hand to help make things clearer.

OBJECTIVES FOR THIS LESSON

Use a model to explore the double-pump action of the heart.

Determine the direction of the flow of blood through the heart.

Recognize that humans have a closed circulatory system.

Study the structure of the human heart.

Explain the differences between pulmonary and systemic circulations.

Identify the strengths and limitations of the siphon-pump heart model.

DR. WILLIAM HARVEY CLOSES THE LOOP

Dr. William Harvey

It was the most important moment of Dr. William Harvey's life. The year was 1616, and he was about to deliver a most surprising message to his colleagues at London's Royal College of Physicians.

On that day, Harvey proposed that humans have a closed circulatory system. Blood recirculates around our bodies, he said. Time after time, the same blood moves out of the heart to the body, back to the heart, out to the lungs, and back to the heart. The blood remains confined within the circulatory system.

What is so surprising about this? Nothing at all if you're living in the 21st century. But Harvey knew that his demonstration would raise some eyebrows. In presenting his hypothesis, he went against 1500 years of medical tradition.

Earlier physicians believed that human bodies were constantly producing enormous quantities of new blood. Harvey had a different idea, and he based his conclusions on direct evidence. He dissected human cadavers and exposed the hearts of living animals. He watched how things worked. His hypothesis that blood recirculates was based on a mathematical calculation.

He began by estimating how much blood was forced out of the heart with each beat. His guess was about 2 ounces. Multiplication took care of the rest. If 2 ounces of blood are ejected with each beat, and the heart beats 72 times a minute, that's 144 ounces of blood a minute. Multiply that by 60, and you get 8640 ounces (about 60 gallons) of blood flowing through the heart every hour!

Sixty gallons of blood would weigh more than 420 pounds. It's obvious that an average human being has far fewer than 60 gallons of blood. For Harvey, the logical conclusion was that blood does not just move from the heart outward to all parts of the body—it circulates back to the heart.

MATERIALS FOR
LESSON 14

For you

1 copy of Student
 Sheet 14.1a:
 Human Circulation
1 copy of Student
 Sheet 14.1b:
 Model Circulation
1 copy of Student
 Sheet 14.1c: Heart
 Structures
1 summary box

For your group

1 tote tray
2 siphon-pumps
2 large plastic cups
1 black marker
1 piece of newsprint
1 set of colored
 pencils
 Water (or access
 to a sink)

Getting Started

1. Listen as your teacher talks briefly about the bottles that are sitting on the desk at the front of the room.

2. Discuss with the class the important discovery that Dr. William Harvey made about the human circulatory system.

3. Discuss the structure of the heart and the circulatory system briefly with your group. Then have one member capture your ideas on newsprint in the form of a drawing. Label any parts you think you know. Indicate on the drawing how you think blood flows through the heart and body.

4. Have someone from your group share your illustration with the class.

5. Put your names on the newsprint and give it to your teacher.

Figure 14.1 *This pump will help you understand how another pump—the human heart—works.*

Inquiry 14.1
Analyzing the Siphon-Pump Heart Model

PROCEDURE

1. You will work with your group for this lesson. Have two members from your group pick up the materials. One person should carry the tote tray, the other the two cups of water. Put the cups in the tote tray after you return to your desk.

2. Use one siphon-pump and one cup of water to explore how the pump works (see Figure 14.1). Make sure the ends of both tubes remain in or pointed toward the water in the cup.

3. Discuss the following questions with your group and record your answers in your science notebook.

   A. What makes the water begin to flow through the pump?

   B. Through which tube, the stiff one or the flexible one, does water enter the pump?

   C. What keeps the water from flowing back through the tube when you release the bulb?

   D. Listen closely as you squeeze the bulb. Can you hear any clicking sounds? If so, what seems to be causing them?

4. Now work with your group to create a "closed" circulatory system. Use both pumps and containers of water to create your system. You must connect the parts so that water flows continuously through the model. You also must work within two constraints: (1) you cannot add water to the cups; and (2) approximately the same amount of water must remain in each cup at all times.

5. When you have a model that you think works correctly, show it to your teacher. Your teacher will then give you a copy of Student Sheet 14.1b: Model Circulation. On it, draw arrows that show the path of water through all parts of the model. Make sure you draw enough arrows to show the entire path.

6. Your teacher will now give you Student Sheet 14.1a: Human Circulation and Student Sheet 14.1c: Heart Structures, and will display a transparency of the heart that appears on Student Sheet 14.1c. Copy the names of the heart structures in the appropriate blanks on the student sheet. Your teacher will discuss some of these structures and their functions.

7. Place Student Sheets 14.1a and 14.1b side by side. Then turn Student Sheet 14.1b upside down. This will make your comparison easier, because the siphon-pump model is actually like an upside-down version of the human heart. Keep the two figures side by side and do the following:

A. Color-code the following comparable structures on both illustrations. Use the following colors for each structure:

Left atrium: blue
Left ventricle: red
Right atrium: brown
Right ventricle: black
Human body: yellow
Valves: green (for all four)
Aorta: orange
Lungs: purple

B. As you did on Student Sheet 14.1b, draw arrows on the figure on Student Sheet 14.1a to show the path that blood takes as it circulates through the human body.

8. Answer the following questions in your science notebook:

A. When you are operating your model heart, you should be able to hear the familiar "lub-dub" sound as well as see what causes it. Explain what causes this sound.

B. Why is the heart considered a double pump?

C. What is the function of the valves in the heart?

D. What do we mean when we say humans have a "closed" circulatory system?

9. Follow your teacher's directions for cleanup.

MARCELLO MALPIGHI— MAN WITH A MICROSCOPE

Dr. Marcello Malpighi

Microscopes had not yet been invented during the lifetime of William Harvey. But fewer than 50 years later, when the Italian physician Marcello Malpighi was beginning his career, this wonderful new scientific tool was available for use in research.

Malpighi studied many human tissues under a microscope. He studied the liver, kidneys, skin, and even the brain.

Malpighi was also interested in the circulatory system. In 1661, he described the network of capillaries that connect the arteries with the veins. This discovery completed the earlier work of William Harvey.

REFLECTING ON WHAT YOU'VE DONE

1. Work with your teacher to discuss your answers to the questions in Step 8 of the Procedure, other aspects of the inquiry, and your reading.

2. Take a second look at the illustration you made at the beginning of this lesson. With your group, discuss what you would need to do to make your illustration more accurate.

3. In your science notebook, make a list of the strengths and limitations of the siphon-pump heart model. Use the same format that you used when you assessed the bell jar model of breathing in Lesson 10. Discuss your ideas with the class.

BACK IN CIRCULATION

Agents Peppi and Bollo have filed a report to headquarters and gotten the go-ahead to continue their mission.

"Today, we're going to begin to explore the human circulatory system," says Peppi. "You're going to learn a lot of new things. You're also going to have the chance to apply some things you've already learned. This is because one of the remarkable things about human body systems is that everything is related. The more you understand one system, the better prepared you are to explore the next one."

"Sounds good to me," says Bollo. "Where's the door?"

"No door!" exclaims

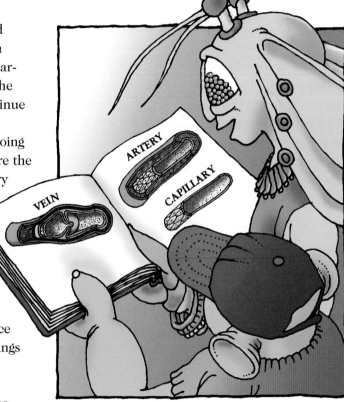

Bollo takes a look at the three types of blood vessels.

Peppi. "There is no entry or exit point to the circulatory system. No mouth, no nose, no anus. The circulatory system is a *closed* system. It consists of the heart, arteries, veins, capillaries, and blood.

"If something breaks this closed system—for example, if a human gets injured and bleeds heavily—it could mean serious damage. The body needs about 5 liters of blood to stay alive. And that

blood has to stay in its place—within the closed circulatory system. External bleeding, which happens when a human is cut, and internal bleeding, which happens when blood leaks beneath the skin, can be dangerous."

Super Transport

"Some people compare the circulatory system with a transportation system. The heart is the hub. It's an important organ, and the human body is designed to offer it protection. The heart is nestled in the chest cavity, cushioned by the soft, spongy lungs and surrounded by a sac called the pericardium.

"Large, one-way vessels carry blood from the heart. These are the arteries. The arteries branch out into smaller vessels called arterioles. Eventually, the 'passengers'—red and white blood cells as well as cell fragments—pass into the smallest, thinnest-walled vessels along the line."

"The capillaries, right?" asks Bollo, remembering their experience in the respiratory system.

"Right. Once a drop of blood has reached its destination, unloaded oxygen and nutrients, and picked up some carbon dioxide and other waste materials, it's ready to go back to the heart, by way of the veins. The veins, too, are one-way streets.

"It's an amazing transportation system. If placed end to end, the blood vessels in a human adult would be about 96,000 kilometers long—enough to stretch around the Earth two and a half times! It's efficient. The human heart pumps about 7200 liters of blood daily. And the time it takes for one blood cell to enter the heart, move through the body, and return to the heart

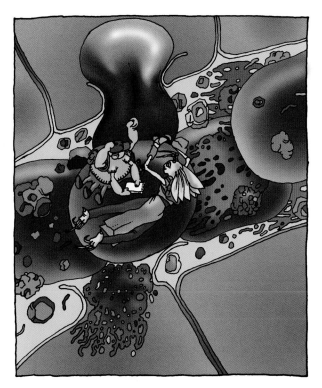

Aboard a red blood cell

is—well, just guess!"

"Five minutes?" says Bollo.

"No. Just 60 seconds when the person is resting," replies Peppi.

"Wow. I'm impressed. Let's go inside and watch this transport system in action," says Bollo.

"Great. To enter, we'll attach ourselves to an oxygen molecule again. All aboard!"

Twin Pumps

"Time for a quiz," says Peppi as they sail down into the lungs. "Do you remember what happened when we traveled on a red

blood cell after leaving the lung?"

"Sure thing."

"You might have expected that we'd head straight for the body cells that needed a fresh supply of oxygen, right? Instead, we went . . ."

"To the heart!" says Bollo.

"Good for you. And after we had delivered our oxygen to the body cells and were ready to go back to the lungs, we went through the heart a second time."

"Yes. I was wondering about that," says Bollo. "Why couldn't

we just head straight back to the lungs? After all, we didn't have any oxygen with us anymore. I was wiped out."

"Keep your eyes open. You'll see," says Peppi.

Peppi and Bollo move through the cells making up the wall of an alveolus and into a capillary. Their oxygen molecule latches on to a red blood cell and heads for the left side of heart.

"Now we're in a pulmonary vein," says Peppi. "The heart is right ahead."

They enter the left atrium of the heart from the back. Squish! The walls of the chamber contract, and they shoot downward through a valve.

They arrive in another chamber. Its walls are much thicker than those of the atrium. Peppi and Bollo take a seat on a muscle fiber at the bottom of the heart.

"Let's stop for a moment," says Peppi. "Now tell me what's happening."

"The heart is beating regularly," says Bollo. "There's a

squeeze at the bottom, then a squeeze at the top. Lub-dub, lub-dub, lub-dub." He checks his stopwatch. "A little more than one beat per second. How does the heart know when to beat?"

"Heartbeats are caused by electrical impulses that come from a small area in the upper right corner of the heart. These impulses make the heart muscles contract. This pocket of specialized muscle cells is called the pacemaker. Some people with heart trouble have to get artificial pacemakers to keep their heart beating properly."

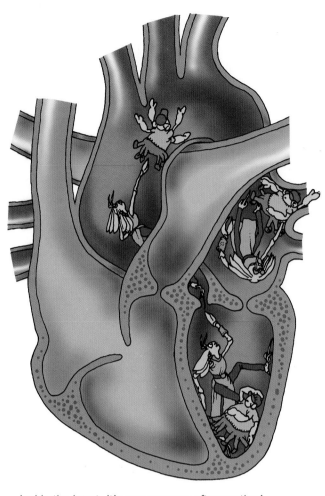

Inside the heart. It's one squeeze after another!

Moving On

The break is over. Whoosh! Blood moves up and out of the heart, carrying Peppi and Bollo with it.

"This must be a major highway," says Bollo. "It's broad. Room for lots of traffic."

"Yes, this is the aorta. It's a super-highway as far as human blood vessels go—about 3.5 centimeters in diameter."

Peppi and Bollo head out via the aorta. It soon divides into smaller arteries that branch out to various parts of the body. The spies go with the flow, heading north to the brain.

Peppi and Bollo watch some oxygen being delivered to the cells of the brain and see some carbon dioxide being picked up in exchange. This all happens through the walls of the capillaries.

They soon find themselves in a vein and heading back to the heart. But this time, they're traveling a different route. They're still entering from the top, but now they're on the right side of the heart. They enter. A moment later, they are being squeezed downward. Another contraction,

and up again.

The journey continues. "We're now in the pulmonary artery," says Peppi. "We're moving out of the heart and back to the lungs."

Four Rooms, Two Pumps

"Hold on. I'm confused," says Bollo. "We've been in four chambers. We started on the left side and we ended on the right. I feel like I've really been shuffled around."

"Here's why. The heart is not just a pump. It's really like two pumps that are side by side. The pumps do different things, but they work 'in synch.' Each of these pumps has an upper chamber, called the atrium, and a lower chamber, called the ventricle. The left and right sides are separated by a thick, muscular wall called the septum.

"The atria (that's the plural of 'atrium') contract at the same time. Blood in both sides of the heart moves through valves to the ventricles. The blood on the left side of the heart has come from the lungs. It's bright red, loaded with oxygen, and ready to go to work. The blood on the right has come from elsewhere in the body. It's done its job. You can tell, because it's got the 'blues.' It's carrying less oxygen and more carbon dioxide and other wastes.

"The ventricles contract. Blood in both sides of the heart moves into arteries. The bright-red blood moves into the aorta. The tired, dark-red blood moves into the pulmonary artery. It's headed for the lungs and a fresh dose of oxygen."

"Two systems in one!" says Bollo.

"Exactly. Humans have pulmonary circulation between the heart and the lungs *and* systemic circulation between the heart and the rest of the body. And the systemic circulation has many smaller divisions, such as those to the liver and kidneys. Quite a big job for an organ that's no bigger than a human fist and that weighs only about half a kilogram," says Peppi.

"Now, guess what!" Peppi continues. "Our voyage through Joanne's circulatory and respiratory systems is over. I'm in the mood for a swim. How about you?"

"Great idea. Just let me get my rubber duck," says Bollo. □

Peppi and Bollo take a dip.

Factors Affecting Heart Rate

How would a roller coaster ride affect your heart rate? Why?

INTRODUCTION

Have you ever been so scared that you could feel your chest pounding? Has your mom or dad ever complained of a "throbbing" headache? Those pounding and throbbing sensations are both caused by the beating of your heart.

Heart rate is among the many body functions that are "on automatic." Your heart rate slows down or speeds up, depending on what your body needs at the moment.

In this lesson, you will learn how to measure your heart rate by feeling your pulse. First, you will record your heart rate when you are sitting at your desk. This is called your resting heart rate. Then you will see whether your heart rate changes under different conditions. You will design your own inquiry, from start to finish. When you have completed your inquiry, you will share your results with the class.

OBJECTIVES FOR THIS LESSON

Measure resting heart rate by feeling your pulse.

Design and perform an inquiry that explores a factor that may affect heart rate.

Identify factors that may cause changes in heart rate.

THE BEAT GOES ON

When you have a physical examination, your doctor checks your vital signs. Vital signs include heart rate, blood pressure, body temperature, and respiratory rate. Heart rate and blood pressure are indicators of how well your circulatory system is operating. Your pulse is an indication of your heart rate.

What exactly is a pulse? Suppose that you and your lab partner were each holding the end of a rope. All of a sudden, you snapped your end of the rope. You would see a wave travel the length of the rope. Your partner would be able to feel the wave when it reached his or her hand.

This is somewhat similar to what happens when blood travels from your heart through the arteries. The "snap" is caused by the blood that is pushed through the arteries by the contractions of the ventricles. The snap occurs at regular intervals. You can feel a "wave" with your fingers when you press an artery against firm tissue (for example, at your inner wrist, where the radial artery passes) over bone. That wave is your pulse. To measure your heart rate, all you need to do is count how many pulses, or beats, occur in 1 minute.

How to feel the pulse at the wrist

You can also feel a pulse at the carotid artery in the neck, the temporal artery at the side of your forehead, and the brachial artery at the crook of the arm.

Some people have difficulty detecting their pulse. For example, it's harder to detect a pulse in a child than in an adult. In this case, one can use a stethoscope to listen to a child's actual heartbeat.

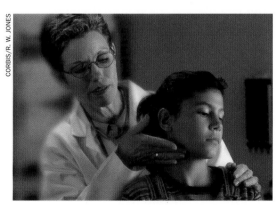

This doctor is checking her patient's pulse by feeling the carotid artery.

MATERIALS FOR
LESSON 15

For you

1 copy of Student Sheet 15.1: Effect of Weight and Exercise on Heart Rate

For your group

1 tote tray
4 1-gallon plastic containers, filled with water and capped (if your group is investigating the effect of added weight)
2 stopwatches
4 rulers

Getting Started

1. Read "The Beat Goes On," on page 131. Then practice with your group until each of you is able to detect your own pulse and that of your partner. Then, after sitting quietly in your seat for at least 1 minute, measure your heart rate for 1 minute. Record this number in your science notebook as your "Resting Heart Rate."

Your heart rate is expressed in beats per minute. Does that mean you need to count the beats for a whole minute?

2. When your teacher calls your name, state your heart rate. Your teacher will record each heart rate while a student volunteer totals them.

A. How would you determine the average resting heart rate for all the students in your class?

B. What is the average resting heart rate of the students in your class?

C. How does your heart rate compare with the class average?

3. Now that you have measured your resting heart rate, work with your group to develop a list of things that you think could make your heart rate lower or higher. Record your list in your science notebook. Your teacher will record responses from each group on a class list.

Inquiry 15.1
Exploring Factors That Affect Heart Rate

PROCEDURE

1. Many things can affect heart rate. In this lesson, you and your classmates will explore two of these factors—additional weight and exercise. Your teacher will tell you which factor your group will investigate. You will work with your group to design the inquiry. Then you will split into pairs to carry it out.

A. If you are going to explore the effect of increased *weight* on heart rate, each pair in your group will need two 1-gallon plastic containers filled with water. The two containers combined weigh about 6.8 kilograms.

B. If you are going to explore the effect of *exercise* on heart rate, you may choose the exercise you wish to do, but you will need to get your teacher's approval before you start. The exercise should be something that can be done safely in the classroom. If you can go outdoors, your choices will be greater.

2. You will use a stopwatch to time your partner's heart rate.

3. Refer to Part 1 on Student Sheet 15.1 to design your inquiry and record your results. Note that you will need to include information such as the following:

Question(s) I will try to answer
Materials I will use
Procedures I will follow

My data table
Graph of my findings
What I found out

4. Be sure to set up a control for your inquiry.

5. Design a data table on your student sheet. Because all members of your group will share data, your table must allow space for each person in the group. Give the table a title.

6. Use the graph at the end of Part 1 on the student sheet to display your group members' heart rates under the conditions explored in your inquiry. Label your graph with the appropriate units and give it a title.

7. When you have finished your investigation and have completed Part 1 of the student sheet, answer Part 2, which asks you to predict daily activities that may cause a rise or fall in heart rate.

Tennis champion Bjorn Borg, a five-time winner at Wimbledon, had a resting heart rate that was just about half that of an average adult. What does that indicate about his heart?

REFLECTING ON WHAT YOU'VE DONE

1. With the class, discuss "Blood: Life's Liquid."

2. Discuss the results of your inquiry.

3. Look again at the list you made in your science notebook during the "Getting Started" activity. Revise it if necessary.

4. You have now learned something about the effect of exercise on heart rate. How do you think regular exercise might affect heart rate?

5. You have also learned about the effect of increased weight on heart rate. The increased weight, however, was just temporary. What do you think would be the effect of a permanent weight gain on heart rate?

6. What factors do you think might affect your heart rate in the course of a normal day? Share the information you recorded in Part 2 of your student sheet, about things a person could be doing at the times listed. Be prepared to defend your responses.

BLOOD: LIFE'S LIQUID

If you drained out all your blood (yuk!), you could fill a 3-liter bucket. Adults have even more blood than young persons do. An adult man of average size (around 70 kilograms) has about 5 liters of blood in his body; a grown woman who weighs around 50 kilograms has about 3.5 liters of blood.

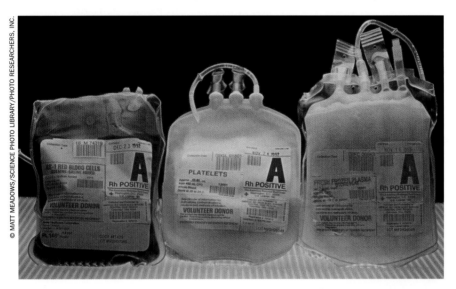

What blood components do you see named on these bags?

Every cell in your body depends on blood to deliver food and oxygen and to carry away wastes. Blood is the key to keeping all the cells of your body alive.

Although it may look like a plain red fluid, blood is a complex blend of liquid and solids. What's in this complicated mix? Only about 55 percent of blood is liquid. This liquid part is called plasma. It is a pale yellow fluid that is 95 percent water. The rest of the plasma is a mixture of dissolved materials, such as sugar, protein, minerals, and wastes that are being transported to and from your body's cells.

The solid parts of blood are trillions of red cells, white cells, and platelets. They ride along in the plasma and are responsible for feeding and protecting the body.

Red Cells—Blood's Work Force

Oxygen is one of the most important substances in the blood. The cells that transport oxygen are called red blood cells (erythrocytes). An adult man has about 25 trillion red blood cells. The bright-red color of these cells develops when oxygen combines with hemoglobin inside the red blood cells. Hemoglobin contains iron, and it attracts oxygen like a magnet. Red blood cells are shaped like disks that have been squeezed in the center. Each red blood cell can carry almost a billion oxygen atoms!

As the blood travels through your body, it delivers oxygen and nutrients to the cells. Blood that has traded its oxygen for carbon dioxide is not bright red anymore. It's

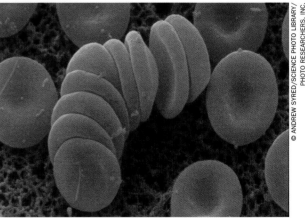

Red blood cells are the most common type of cell in human blood. They are seen here under extremely high magnification using a scanning electron microscope. Can you understand why some people describe red blood cells as "doughnuts without holes"?

dark red or purple. That's one reason why the veins near the surface of your skin look so dark.

At any one moment, you have enough oxygen in your red blood cells to keep you alive for only about 5 minutes. Red blood cells work nonstop—to keep oxygen moving from the air you breathe to the cells that need oxygen and nutrients. It's hard work. About 2 million of a person's red blood cells die every second. But don't worry. Your body is constantly making new ones in the marrow (the soft center) of flat bones such as the hips, ribs, and skull.

White Cells: The Warriors

White blood cells (leukocytes) are your body's warriors. They're part of your immune system. White cells are bigger than red cells, but red cells outnumber white cells by about 650 to 1. White cells can change shape and squeeze between other cells to patrol your whole body, looking for invaders such as bacteria or viruses.

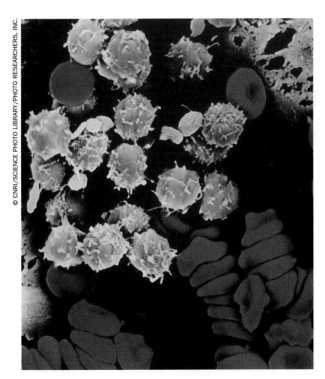

This electron micrograph shows immature red and white blood cells in bone marrow. Because cells live only a short time, the bone marrow must constantly produce new ones.

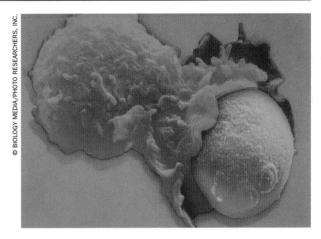

This electron micrograph shows a white blood cell engulfing a yeast cell.

When the white blood cells find an invader, they attack. Some white cells surround and digest the microbes. Others produce chemicals that kill or cripple the enemy. The white blood cells also produce chemicals that act like an alarm signal and cause your body to send more germ-fighting cells to the battlefield.

Many white cells die in their war against bacteria. The sticky pus that collects around an infection is a pool of dead white blood cells and bacteria. But again, there's no cause for concern. Your body is always producing more white blood cells. In fact, if there are more germs than usual in your body, it produces extra white cells.

Sometimes the body produces too many white cells. A kind of cancer called leukemia causes the body to produce so many white cells that they can outnumber the red cells.

Platelets: The Repair Crew

Because blood is so important, your body has a way to patch holes in the blood vessel walls so blood doesn't leak out. This activity, called clotting, is done in part by platelets. When your skin is cut, the platelets are exposed to air and they begin to fall apart. This releases chemicals that react with a series of compounds in the blood (one of which is fibrinogen) to start forming a web of tiny threads. These threads twist together to form a web that traps red blood cells. The blood cells dry out and form a scab.

Clots can form inside your body, too. The black-and-blue marks that you call bruises are really clots underneath your skin. Your body needs calcium and vitamin K to help form some of the chemicals that help control clot formation. If you don't have enough of these nutrients, it will take a long time for your blood to clot.

Some people are born with a disease called hemophilia that prevents clots from forming at all. For them, even a small scratch can be dangerous, because they might lose a great deal of blood.

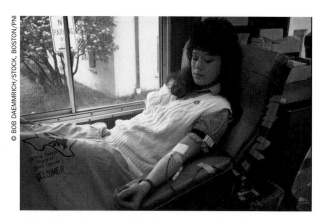

Giving blood is not painful. Many people do it on a routine basis. Some communities have "bloodmobiles" that travel to workplaces, schools, and other public places so that it is easier to give blood. This college student is giving blood at a bloodmobile.

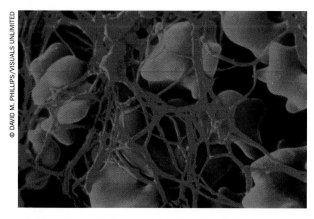

Clotting blood. Notice how the threads of fibrinogen have trapped the blood. If you look closely at a scab, you can see evidence of these threads (magnification x 5000).

Blood Types: Not All Blood Is the Same

If you lose a lot of blood because you've been in an accident or had an operation, you may need a transfusion. This means that blood is transferred into you through a thin tube that has been inserted into one of your veins. But it has to be the right type of blood.

There are four main types of human blood. An Austrian scientist named Karl Landsteiner discovered the different blood types in 1901. Dr. Landsteiner named the four blood types A, B, AB, and O. The letters tell what kind of chemical markers are found on the surface of red blood cells. Type A blood has one kind of marker; type B has another. People with type AB blood have both markers. Type O blood has neither A nor B markers.

Dr. Landsteiner also discovered another blood type called the "Rh factor." It was named after the rhesus monkey in which it was first found. People whose blood has the Rh factor are referred to as "Rh positive." Those who lack this factor are "Rh negative." People inherit this blood type independently of the other types; therefore, we usually describe a person's blood type by including both the type and the factor

The Rh factor in human blood is named after the Rhesus monkey, in which the factor was first detected in 1942.

(for example, "O positive" or "AB negative").

In their plasma, people have antibodies against the red blood cell markers that they *do not* have. If someone receives a transfusion of a blood type that contains markers other than their own, antibodies will attack the foreign red blood cells and cause them to stick together. This can result in serious damage, or even death, because the clumps of cells can clog small vessels.

Do you know your blood type? If so, check below to see what types you can donate to and receive blood from safely. ☐

Blood Types, Donors, and Recipients

Blood Type	Can Donate Blood to	Can Receive Blood From
A	A, AB	A, O
B	B, AB	B, O
AB	AB	A, B, AB, O
O	A, B, AB, O	O

QUESTIONS

1. Study the table. Why do you think that type O blood is sometimes called the universal donor?

2. Bearing in mind that type O blood is called the universal donor, what might be a good name for type AB?

3. If you have blood type O, what blood type or types may you receive safely?

The Other Circulatory System

Did you know that you have another circulatory system? It is called the lymphatic system. When blood flows through the capillaries, some of it leaks out and squeezes between the body cells. It is then called tissue fluid. Tissue fluid is an important link between the blood and the body cells. It helps the exchange of wastes and nutrients occur more smoothly.

Tissue fluid either returns to the bloodstream through the capillary walls or through another system of vessels called lymph vessels.

The lymph vessels pass through lymph nodes, which are tiny swellings that filter the tissue fluid as it passes through. The lymph nodes are always on the lookout for invading germs, and the minute they detect a foreign invader, they get to work. The lymph nodes contain cells that either fight bacteria directly (by "eating" them) or by producing antibodies that destroy the bacteria.

The largest collection of lymph nodes are in the armpits, neck, and groin. When your doctor feels your neck during an examination, he or she is trying to see whether your lymph nodes are swollen. This might indicate that you have an infection. If your body's own defense mechanism (those white cells and antibodies) isn't strong enough to do the job, the doctor might have to prescribe an antibiotic.

Lymph is also important because it can carry some nutrients that are too large to enter the bloodstream through the cells of the capillary walls. For example, the lymph system carries digested fats to some of the large veins. The veins then take the fats to the liver, where they are processed.

The lymphatic system provides another good example of the teamwork needed to keep the human body healthy.

The Heart Meets Resistance

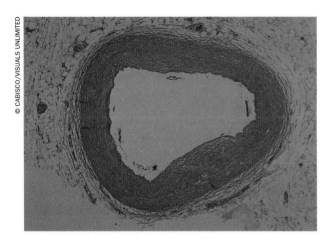

Cross-section of a healthy human artery. Note the wide opening in the middle.

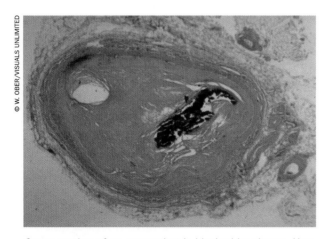

Cross-section of an artery that is blocked by plaque. How do you think the blockage would affect the heart's ability to pump blood throughout the body?

INTRODUCTION

You have explored and read about the structure and function of the heart and blood vessels. As a result, you have a good idea of what happens when the circulatory system is working well.

But what if something goes wrong? For example, what if an artery becomes partly blocked? Does it affect the heart?

In this lesson, you will investigate what conditions can make it harder for the heart to pump blood through the body. You will be exploring blood pressure, which is an important indicator of how well the circulatory system is working.

As you begin to explore blood pressure, it might help to think about two hoses that are connected to a water pump. The diameter of one hose is twice as wide as that of the other. Would there be any differences in the pressure of the water that flows through the two hoses? You will soon find out!

OBJECTIVES FOR THIS LESSON

Use a model to explore whether the diameter of a tube affects how hard a pump must work to circulate water through it.

Examine the burden that narrowed or hardened arteries place on the heart.

Getting Started

1. With your group, list in your science notebook everything you know about blood pressure and the things that you think can make it higher or lower.

2. Your teacher will record your ideas on newsprint.

Inquiry 16.1
Feeling the Pressure

PROCEDURE

1. Have someone pick up a tote tray for your group. You will work in pairs for this inquiry.

2. Listen as your teacher describes how to perform the inquiry. Then construct a data table in your science notebook.

MATERIALS FOR
LESSON 16

For you
1 copy of Student
  Sheet 16.1: Study
  Guide—Respiration
  and Circulation

For your group
1 tote tray
2 siphon-pumps
2 large plastic cups
2 rubber stoppers
  with large holes
2 rubber stoppers
  with small holes
2 stopwatches

3. Conduct the inquiry as follows:

A. Hold the siphon-pump in a vertical position. Place the end of the tube that extends straight down from the bulb close to the bottom of the cup.

B. Hold the end of the tube that is attached at the side of the siphon-pump just above the surface of the water in the cup, as shown in Figure 16.1. Make sure the tube is pointed toward the water. This tube, which does not have a stopper, represents a normal artery.

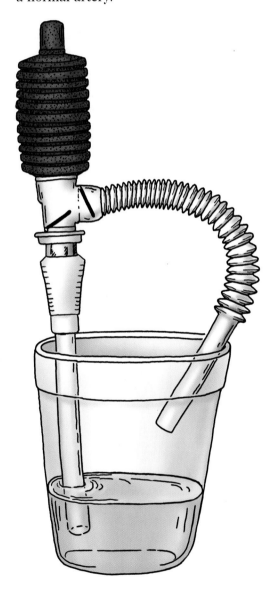

Figure 16.1 *How to position the pump within the cup*

C. Grip the pump between your thumb and first three fingers. While your partner times you with a stopwatch, see how many times you can squeeze the pump in 15 seconds. Make sure that your thumb and fingers meet through the bulb each time you squeeze. Let the bulb return to its original position before squeezing again.

D. Have your partner record on your data table the number of squeezes you make.

E. Switch roles with your partner and repeat the activity.

F. Repeat steps A through E. Then determine the average number of squeezes you made in 15 seconds. Record the average on your table.

4. Now insert the stopper with the large hole into the end of the tube that is attached to the side of the siphon-pump. Repeat Steps 3A through 3F. Both you and your partner should perform two sets of squeezes under each condition. Record your data.

5. Remove the stopper with the large hole and insert the stopper with the small hole. Repeat Steps 3A through 3F. Again, record your data.

6. Empty the water into a sink or container. Place the cup and siphon-pump in your tote tray and return the tray to the designated area.

7. Refer to the reading selection "Blood Pressure: What Goes Up Should Come Down," on page 142, as well as the results of your inquiry, to help you answer the following questions in your science notebook:

A. Under what conditions were you able to complete the most squeezes in 15 seconds? Explain.

B. Under what conditions did the model represent the heart pumping into an artery that was heavily lined with plaque?

C. Write one sentence in your science notebook describing how your hand felt after completing the exercise under each condition.

D. On the basis of how your hand felt during the test with the small-holed stopper, what do you think might happen to your heart if it had to pump blood through an artery whose diameter was narrowed by plaque?

E. On the basis of what you have discovered in this lesson, what do you think it means if someone has "high" blood pressure?

F. Figure 16.2 depicts an electronic reading of someone's blood pressure. Why do you think the line goes up and down repeatedly? List two things that would cause the line to go higher.

REFLECTING ON WHAT YOU'VE DONE

1. Discuss with your class the results of your inquiry. Can you now answer the question about the pump and hoses at the end of the Introduction to this lesson?

2. Take a second look at the list you developed in "Getting Started." Revise your list if necessary.

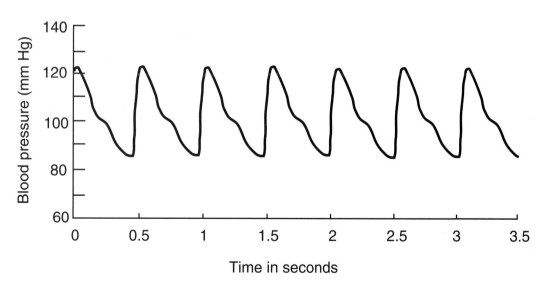

Figure 16.2 *Electronic blood pressure reading*

BLOOD PRESSURE:
WHAT GOES UP SHOULD COME DOWN

You already know that your heart works like a pump. It pushes blood through your

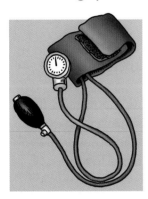

A blood pressure cuff, or sphygmomanometer

blood vessels much like a pump pushes water through a hose. Both pumps have to work against resistance. If the hose or blood vessels have narrow openings, there is more resistance, and the pump or heart must work harder.

Your blood pressure is an indication of how much resistance your heart must overcome to keep blood flowing through your blood vessels.

You've probably had your blood pressure measured at the doctor's office. Blood pressure is so important that it is a routine part of any physical examination. Your doctor uses a blood pressure cuff to measure the blood pressure in your large arteries—usually those of the arm. The doctor wraps the cuff around your upper arm. Then she uses a squeeze bulb to pump air into the cuff. The cuff expands and presses against an artery in your arm until blood stops flowing through it. A gauge shows how much pressure is in the cuff. The air in the cuff is then slowly released. When blood begins to move through the artery again, a faint tapping sound can be heard with the stethoscope. The pressure indicated on the gauge at this moment is known as the systolic pressure.

Then the doctor lets air out of the cuff until blood starts flowing normally and the tapping sounds can no longer be heard. The pressure on the gauge at this moment is known as the diastolic pressure. The two measurements are expressed as a fraction, for example, 110/60. The systolic pressure is the numerator, and the diastolic pressure is the denominator.

CORBIS/RICHARD T. NOWITZ

Ice cream is great, but it's high in fat. Why not try yogurt or fruit instead?

As plaque builds up in an artery, the diameter of the artery narrows.

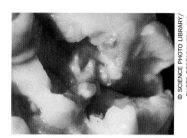

© SCIENCE PHOTO LIBRARY/
PHOTO RESEARCHERS, INC.

Plaque that has been removed
from a patient's arteries

The average blood pressure reading for a healthy young adult is 120/80. An adolescent's blood pressure is lower—about 100/65 for a 13-year-old boy or girl.

If your blood pressure is too high, it can cause health problems. High blood pressure, or hypertension, places a burden on your heart. This can weaken your heart. High blood pressure is also dangerous for your blood vessels. It may cause a vessel wall to break.

Doctors can't explain all the causes of high blood pressure. Some illnesses can cause this condition. Heredity plays a part. So does diet. Too much salt, for example, is a risk factor for high blood pressure. Cholesterol (a kind of fat) in the food you eat can get deposited as plaque inside your blood vessels and cause the vessels to narrow. This may cause high blood pressure and, perhaps eventually, a heart attack.

If your blood pressure is too high, your doctor might prescribe medication. She might also recommend that you change your diet and get more exercise.

High blood pressure often doesn't have any symptoms. Sometimes it's called the "silent killer." That's why it's important to measure your blood pressure regularly, even when you are young. You can find stations where you can measure your blood pressure free at pharmacies and supermarkets. Take advantage of them! You can also buy blood pressure measuring devices to use at home. ☐

Cholesterol: Friend or Foe?

Cholesterol is a soft, waxy substance that is found in all the cells of your body. It is made in the liver. It is used to form a part of cell membranes and other tissues.

Everyone needs some cholesterol for normal body functions. But too much cholesterol and fat can attach to the walls of blood vessels and cause plaque to build up. The plaque narrows the arteries, and the heart has to work harder.

The extra burden on the heart can lead to heart attacks. Or, if a heart artery (called a coronary artery) becomes blocked, the cells that are nourished by that artery die. If a sufficient quantity of heart cells die, a heart attack may occur.

Controlling cholesterol can be difficult. For one thing, heredity plays a role in determining cholesterol levels. What's more, scientists disagree on how to control the cholesterol levels in our blood. Medications work for some people; for others, different approaches, such changes in diet, are helpful.

Scientists do agree that you can do several things to reduce your chances of heart disease. These include engaging in regular exercise, avoiding tobacco products, maintaining a healthy weight, and identifying and treating high blood pressure.

17

The Respiratory and Circulatory Systems—An Assessment

The first part of this assessment will require some help from your body as well as your brain!

INTRODUCTION

During the past seven lessons, you have seen how the respiratory and circulatory systems work. You've also learned that these two systems are interdependent.

Now it's time to see what you've learned! This lesson will take two or three class periods. In Part A, you will design and conduct an investigation to determine the effect of exercise on breathing rate. For Part B, you will answer some written questions and complete the revision of your human body systems poster.

When you have completed this assessment, you will be ready to begin the final part of this module, which concerns the musculoskeletal system.

OBJECTIVES FOR THIS LESSON

Complete a two-part assessment on the respiratory and circulatory systems.

Reinforce what you have learned about the respiratory and circulatory systems.

Complete your group's human body systems poster.

Assessment Part A
Designing and Conducting an Inquiry

1. For Part A of your assessment, you will work with a partner to design and conduct the inquiry described on Student Sheet 17.1a. The purpose of the inquiry is to determine the effect of exercise on breathing rate. You will design and carry out the inquiry in one 45- or 50-minute period, so staying on task is a must!

2. After your teacher has reviewed the Procedure, work with your partner to design your inquiry. When you have completed your design, obtain a stopwatch from your teacher and conduct your inquiry.

3. When you have finished conducting your inquiry, double-check to make sure that you have completed Student Sheet 17.1a, including the graph. Use the Inquiry Scoring Rubric (page 63) and the Inquiry Checklist (page 63) from Lesson 8 as guides to make sure your work is complete.

4. Follow your teacher's directions for turning in your work.

MATERIALS FOR LESSON 17

For you

1 copy of Student Sheet 17.1a: Respiratory and Circulatory Systems: An Assessment, Part A—Designing and Conducting an Inquiry

1 copy of Student Sheet 17.1b: Respiratory and Circulatory Systems: An Assessment, Part B—Selected-Response Items

1 copy of Student Sheet 17.1c: Answer Sheet, Part B—Selected-Response Items

For you and your partner

1 stopwatch

For your group

1 tote tray
1 human body systems poster
1 black marker
1 roll of clear tape
1 pair of scissors
1 eraser
Summary boxes

Assessment Part B
Completing Selected-Response Items and Final Revision of Human Body Systems Posters

1. You will work individually on Part B of your assessment. Follow along on Student Sheet 17.1b while your teacher explains this part of the assessment.

2. Your teacher will give you an answer sheet (Student Sheet 17.1c) and directions for completing this assessment. Note that you must wait until all members of your group have turned in their answer sheets before your group can begin to work on the human body systems poster.

3. With your group, complete the revision of your human body systems poster in the following manner:

A. Place any misplaced organs in the correct positions.

B. Attach each summary next to the appropriate organ.

C. Give the poster to your teacher.

REFLECTING ON WHAT YOU'VE DONE

1. Your teacher will return Student Sheet 17.1a and your answer sheet for Part B. Referring to the transparency that shows the questions on Part B, discuss the assessment. Ask questions about anything you still do not understand about the respiratory and circulatory systems.

2. With the class, discuss "Organ and Tissue Transplantation: A 'Round-the-Clock Need," and develop a list of the pros and cons of organ and tissue transplantation.

Organ and Tissue Transplantation: A 'Round-the-Clock Need

Every day, many people in this country receive a new chance for life because of transplantation surgery. Typical transplant patients might include a 60-year-old man whose lungs have failed or a young mother whose kidneys no longer work. Organ recipients also include children.

Surgeons can transplant the human heart, pancreas, liver, kidney, and lung. They can also transplant tissues such as skin, the cornea of the eye, tendons, heart valves, and bone.

The first successful organ transplantation was performed in 1954. Since then, surgical procedures have improved. Scientists have developed better techniques to match the blood and tissue of the donor with those of the recipient. They have discovered new drugs that help increase the chances that the body will not reject the new organ.

Despite this progress, the number of patients who need transplants in the United States is much greater than the number of organ donors.

Nearly 60,000 people are waiting for an organ transplant. Hundreds of thousands more need a tissue transplant.

Every day, 12 people die while waiting for an organ transplant. A new name is added to the transplant waiting list every 16 minutes.

Organs and tissues from a single donor could help up to 50 people in need.

The demand for organs and tissues far exceeds the supply. What's the solution? Find more donors!

People need to know

Once a suitable donor organ is found, it is shipped to the site where it is needed as quickly as possible. That doesn't always require a helicopter trip! This organ transport nurse works at a hospital in Stanford, California.

how easy it is to register to become a donor. Most important, they need to know how many lives they could touch by deciding to become an organ donor. The process involves signing an agreement that when they die, their organs should be available for needy recipients.

What do you think could be done to increase the number of organ and tissue donations and to save many lives? Discuss your ideas with your class. ☐

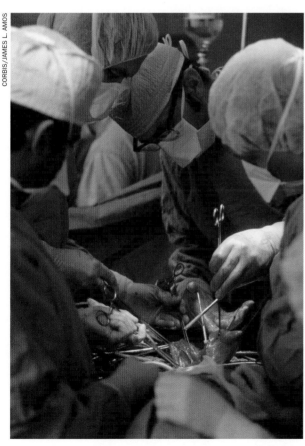

Organ transplantation surgery is a complex procedure. For some organs today, however, the success rate is as high as 95 percent.

PART **3** The Musculoskeletal System

18

The Musculoskeletal System—
An Overview

CORBIS/KIT KITTLE

*Learning to walk is a milestone in life, and no wonder.
It's a big job. Learning to balance and use muscles takes
a good deal of practice and support.*

INTRODUCTION

From the moment they are born, humans are active creatures. Newborn infants wave their arms and legs. Soon, they learn to roll over. The next big challenge is learning to crawl. And once babies take their first steps, they are really on the move! Movement is one of the most important of all human functions. It is also one thing that distinguishes us from most plants.

Like other humans, you are capable of an amazing variety of movements. You can dive into a pool or do the latest dance step. You can also move specific parts of your body whenever you want to. For example, you can raise your hand in class when you think you've got the answer (or duck to avoid the teacher's eyes when you know you don't).

In this part of the *Human Body Systems* module, you will explore how your muscles and bones, with the help of your nerves, use the energy released within your cells to produce movement. You'll begin by focusing on your arm and by comparing it with, of all things, a chicken wing!

OBJECTIVES FOR THIS LESSON

Explore how muscles and bones interact to produce movement.

Reproduce some of the movements made by a chicken when it flaps the lower section of its wing.

Investigate and identify tissues of the musculoskeletal system.

Getting Started

1. Follow along in the Student Guide as a classmate reads aloud the Introduction to this lesson.

2. Now stand up. Close your eyes and stand on your tiptoes. As you do, focus on what your muscles and bones are doing to help keep you erect. Open your eyes. Then close them again and repeat this exercise. As you do so, think about the following questions:

A. What appears to be the role of muscles and bones in helping you keep your balance?

B. What did you notice about how muscles work?

C. What happened when you closed your eyes? What role do your eyes have in balance?

For you
1 copy of Student Sheet 18.1: Arm and Wing
1 pair of safety goggles

For you and your partner
1 dissecting tray
1 chicken wing, in a plastic bag
1 dissecting needle
1 scalpel
1 pair of forceps
2 pairs of gloves
1 small plastic bag
1 black marker
6 paper towels
Water (or access to a sink)

© MICHAEL W. NELSON/STOCK SOUTH/PNI

This ballerina's turns, or pirouettes, require excellent balance. How do her muscles help her keep on her toes?

3. Discuss your responses with the class.

4. Title a blank page of your science note-book "Muscles and Bones." Under the title, draw a line down the middle of the page to make two columns. Title the left column "Muscles" and the right column "Bones." Take 5 minutes to fill the columns with everything you know about muscles and bones.

5. Discuss your list with the class.

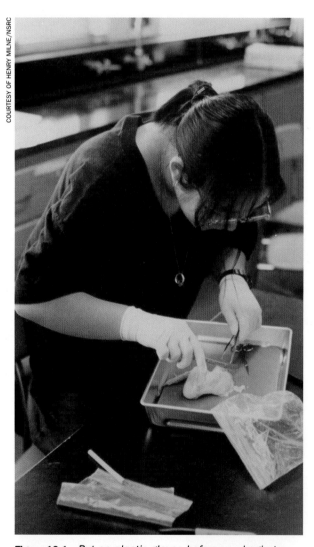

Figure 18.1 *Put on plastic gloves before you begin to dissect the chicken wing.*

Inquiry 18.1
Winging It

PROCEDURE

1. Have one member of your pair collect the dissection tray, a chicken wing in a plastic bag, and the dissecting equipment. With the marker, write your names and class period number on the plastic bag.

2. Listen as your teacher explains the cleanup procedures and reviews the steps in the Procedure for this lesson. If you have a 45- or 50-minute class, you will not finish your dissection in one period. Before the end of class, your teacher will ask you to put your chicken wing back in the plastic bag. You will return the bag to your teacher, who will refrigerate it until the next period.

3. Before you begin, review the Safety Tips with the class.

SAFETY TIPS

Wear plastic gloves and safety goggles during the dissection.

Be very careful handling the blade of the scalpel. Watch as your teacher shows how to tighten it into place. Retract the blade before you return your materials.

Wash your hands thoroughly with antibacterial soap and water after you have cleaned up your work area and returned your materials.

Be careful not to get any disinfectant in your eyes.

4. Your teacher will demonstrate how to remove the skin from the wing. When the demonstration has been completed, put on your goggles and plastic gloves (see Figure 18.1). Follow the directions presented here to dissect the chicken wing and explore what makes it move.

A. Use the scalpel and forceps to remove as much skin from the chicken wing as you can. Pull the skin back gently, as shown in Figure 18.2. Place any waste you cut from the chicken wing in the second plastic bag.

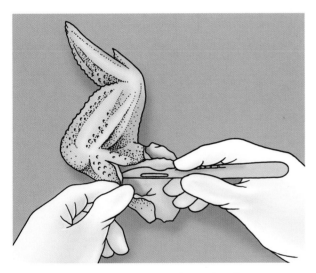

Figure 18.2 *Remove the skin with one hand as you cut away the connective tissue with the scalpel.*

B. You will see a layer of fat right underneath the skin. The fat makes the skin greasy and hard to grip. If you have trouble holding the wing, rinse it with soapy water and blot it dry with a paper towel. Once the skin is pulled back, look at the wing carefully. Answer the following questions in your science notebook:

1. What do you think is the function of the layer of fat underneath the skin?

2. Do you think humans have this fatty layer under their skin?

C. When you have removed as much skin as you can, use Student Sheet 18.1 to identify the following structures:

- muscles (light-brown bundles)
- tendons (silvery-white tissue at the ends of muscles)
- ligaments (cannot be easily seen; they connect bones)
- point of origin (the attached end of muscle that does not move a bone)
- point of insertion (the attached end of muscle that moves a bone)
- joint (where bones meet)
- cartilage (bright-white tissue; softer than bone)
- blood vessels (thin red lines running through the muscles)
- nerve tissue (thin white thread that runs through the muscle)

D. Try to produce a flapping motion of the lower section of the wing. Then answer the following questions in your science notebook.

1. What did you have to do to reproduce the flapping motion of the chicken wing? Describe it in detail.

2. What is the name of the joint that bends and extends to allow this movement?

E. Try to tear apart the upper and lower parts of the wing. Answer the following two questions in your science notebook:

1. Was it hard to tear the sections of the wing apart?

2. What do you think keeps the bones connected?

F. Your teacher will cut one of your chicken bones in half. Look at the inside of the bone and describe in your notebook what you observe. Figure 18.3 illustrates a cross-section of a human bone.

5. Place the remains of the chicken wing and plastic gloves in the plastic bag. Put the bag and dissecting tray in the trash receptacle provided by your teacher.

6. Complete Student Sheet 18.1.

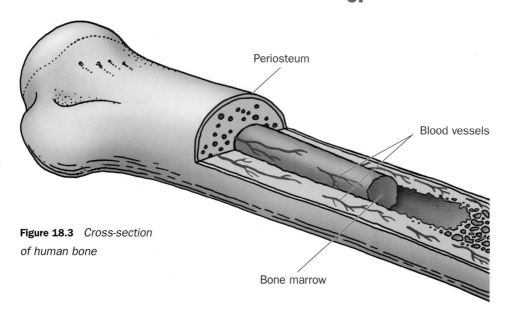

Periosteum

Blood vessels

Figure 18.3 *Cross-section of human bone*

Bone marrow

REFLECTING ON WHAT YOU'VE DONE

1. With your class, discuss the results of your inquiry.

2. Take another look at the list you developed in "Getting Started." Revise it as necessary.

3. If you have not already done so, discuss with the class the questions following "Life in the Bone Zone."

Life in the Bone Zone

What's so great about being a surgeon?

"I can make a difference with my own two hands," says Dr. Laura Tosi.

Dr. Tosi is a pediatric orthopedic surgeon. In other words, she fixes children's bones. Dr. Tosi is chair of pediatric orthopedics at Children's National Medical Center in Washington, D.C.

Dr. Tosi's fascination with bones was kindled when she was finishing her medical training at a New York City hospital. "People would come to the emergency room with fractures, and I could fix them," Dr. Tosi recalls. "It was a great feeling, and I never got over it."

A Miraculous Substance

"Bone is the most miraculous substance in your body," says Dr. Tosi. "For example, it's the only tissue that will heal without a scar. You can have a fracture that looks absolutely awful, but several years later, once it has healed, the bone is just as strong as ever. You can't even see where the fracture occurred."

Some of Dr. Tosi's patients have bone or muscle problems that can't be fixed as easily as broken bones. But there are many ways of helping these children as well. Some undergo surgery to straighten bones or repair muscles. Others benefit from braces. Physical therapy and exercise programs help, too.

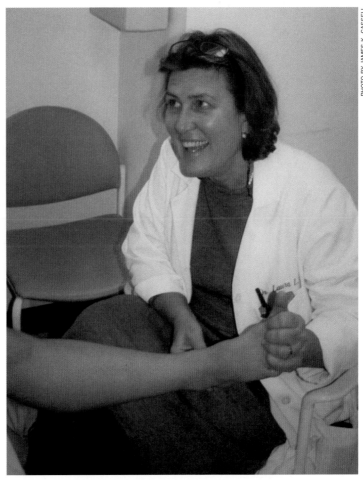

In addition to performing surgery, Dr. Tosi spends many hours in the orthopedic clinic at Children's National Medical Center.

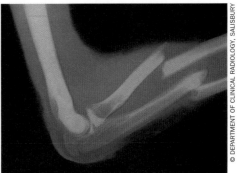

An X-ray of a fracture of the radius (upper right) and ulna (lower right) bones of the human forearm.

These boys aren't just playing basketball—they're building their bone banks.

The "Bone Bank": Time to Invest!

Dr. Tosi is just as interested in building healthy bones as in fixing broken ones. Sometimes it's hard to get her to stop talking about how important it is for children to create a great set of bones while they are still growing.

"Between the ages of 10 and 18, you build most of the skeleton you'll need for the rest of your life. You may live to be 100, but your bone growth is just about complete by the time you're 20," she says. That's why it's important to build a strong "bone bank" when you are young.

How do you build a solid skeleton? "Exercise is critical," Dr. Tosi says. That's why her son and daughter are both active in sports. When Dr. Tosi is on the sidelines, she's not just watching a tennis match or a soccer game. She's cheering for kids who are building their bone banks.

Exercise isn't the only important part of creating a skeleton that will last a lifetime. Diet plays a big part, too. Young people—especially teenagers—need lots of calcium and vitamin D, according to Dr. Tosi.

Where's the best place to get it? "Milk," she declares. "Some people think that because milk has fat, it's bad. But milk is critical to building strong bones." If young people don't drink enough milk, they're not building a bone bank.

They can choose low-fat or skim milk if they're worried about fat.

"Some people," Dr. Tosi adds, "have trouble digesting milk and milk products. They need to make a special effort to get enough calcium. They can get it by eating certain foods, such as soybean products. Calcium-enriched foods and calcium pills also can boost your intake of this important mineral."

Even the strongest young bones can break. Kids fall down. They're injured playing soccer or football. They have accidents. "That will never change," Dr. Tosi says. But one change has reduced the number of broken bones she sees. "Seat belts have made a really neat and wonderful change in my practice. I used to see many children whose bones were broken because they had been thrown around inside a car during an accident. Now that's rare."

Never a Dull Moment

Although her work in the operating room at Children's National Medical Center and its clinics keeps her busy, Dr. Tosi has time for other professional activities as well. She teaches medical students and residents at the hospital. She also does research and publishes articles in medical journals.

Dr. Tosi is also interested in attracting more women to her profession. Although men and women enter medical school in nearly equal numbers today, she says, only three out of

Educating the public about bone health is important to Dr. Tosi. She is pictured with two young friends in front of an educational exhibit displayed at "America Races for Strong Women," which was sponsored by the National Osteoporosis Foundation. Dr. Tosi received an award for her role in promoting this event.

every 100 orthopedic surgeons are women!

Orthopedic surgery is a great career—one that, in Dr. Tosi's view, any young person who's interested in medicine should think about. "Every day, I'm challenged by something new, something hard," she says. "That's what keeps medicine interesting for me. It can be exhausting, but at the end of each day, I know I made a real difference. That's a great feeling." ☐

QUESTIONS

1. Why do you need to build a bone bank when you are young?
2. What are two things young people can do to build a good bone bank?
3. According to Dr. Tosi, what is one thing that provides good protection against broken bones and serious injury?

SPIES

TWO WORKING AS ONE

The Skeleton: Holding It All Together

"Before we begin our adventure inside the human body, let's take a moment to get the big picture," Peppi says. She reaches into her briefcase. "To help us out, we're going to use these special photographs, called X-rays. Doctors on planet Earth use X-rays to see what's going on inside the human body. Usually, X-rays are used to explore a single bone or joint—maybe a broken hip. But here's one that shows

"We have just one more investigation," says Peppi. "It centers on two systems that work so closely together we've got to explore them at once. We're going to look at muscles and bones—the musculoskeletal system."

"If we're going to explore this system in any detail, we're in for a long trip," says Bollo, taking a look at the anatomy guide. "It says here that adults have 206 different bones and more than 650 muscles!"

"Don't worry, we're not going to visit every bone and muscle in the body. We'll just hit some highlights. What we see will give us a good idea of how the human musculoskeletal system works."

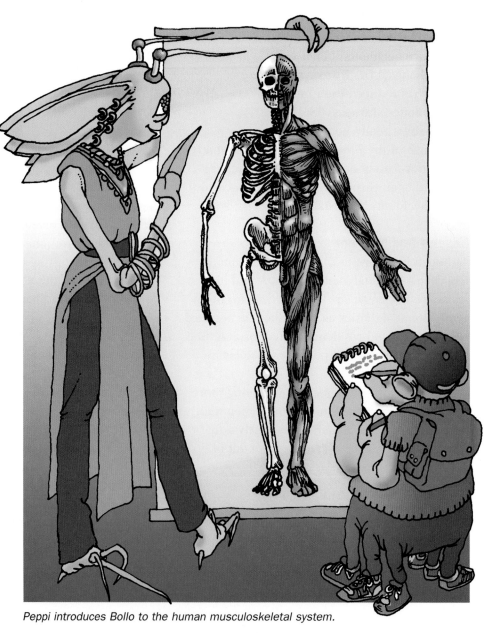

Peppi introduces Bollo to the human musculoskeletal system.

us the entire human skeleton."

"If we didn't already know what humans looked like, this X-ray of the skeleton would be a big help," says Bollo. "A head, two arms, two legs—it's all there."

"Right. And each bone has a special purpose. The tiniest bones are in the middle ear—they are less than 1 centimeter long. But look at the bone in the upper leg. It's the longest bone in the human body. It's the femur, or thigh-bone. In a woman of average height, the femur is usually about 45 centimeters long. The femur is also very strong."

"How strong?" asks Bollo.

"Bone can be as strong as steel in resisting tension," says Peppi. "And the comparison with steel is useful in another way. That's because the skeleton is the body's support system—much like steel girders are the support system for a building."

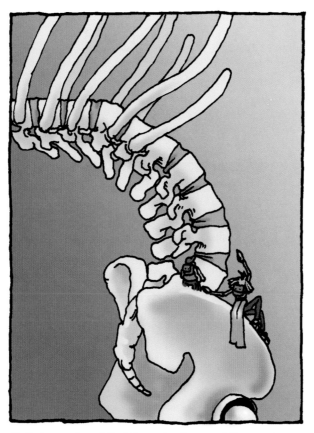

Bollo is convinced that the spine is one of the wonders of the human body. Do you think he is right?

"The backbone looks different from the bones in the legs or arms," says Bollo, looking closer. "It's divided into bumps and spaces."

"Yes. The bones of the arms and legs are straight. Although they are quite strong, they don't bend. That's one reason that bone can fracture. But the spine— the backbone—is a series of 24 small bones, called vertebrae. Because of the discs, which create cartilaginous joints between the vertebrae, the spine is flexible. Humans can 'bend over backwards,' as well touch their toes and perform acrobatics. The discs also act like shock absorbers."

"Interesting!" says Bollo. "I'm beginning to see why humans can be such great athletes. They can jump and withstand pressure. Their bones are strong, yet the joints make them flexible. Anything else I need to know?"

"One more thing. The bones not only support the body but also protect it. Look, for example, at the skull. Pretty solid, right? That's so it can protect one of the human's most valuable organs—the brain."

"And look at the ribs—they surround the heart and lungs, kind of like a cage," says Bollo.

"You've got it," replies Peppi. "Now that we understand the framework, let's move inside."

"Ready when you are," Bollo replies. "I can't wait to see how many of those muscles and joints I can check out for myself." ☐

19

Joints and Movement

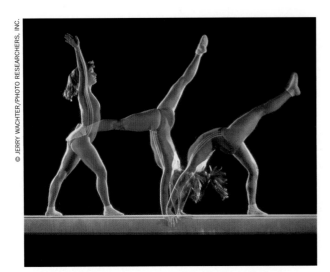

Gymnasts train for years to perform exercises on the beam. Walking on it is one thing, but a front walk-over? There's a real challenge.

INTRODUCTION

Have you ever attended a gymnastics event and been amazed at the performers' contortions? You might be surprised to learn that most of their bending and twisting is just an exaggerated form of the movements that all of us make every day. What enables us to perform these movements? Our joints! Humans have more than 200 bones, and joints are the places where the bones meet. Figures 19.3 through 19.9 (see pages 163–164) illustrate several different kinds of joints.

In this lesson, you will examine the movements that joints allow you to make. You will assemble a model of the arm or of the spinal column. You and your group will use the models to simulate some of the movements human joints make. You also will explore how opposing muscles enable you to move your arm.

OBJECTIVES FOR THIS LESSON

Construct a model of the human arm or spinal column.

Use models to understand the anatomy of the human arm and spinal column.

Use models to explore the form and function of several types of joints.

Identify some of the major joints of the body and describe the types of movements they allow humans to make.

Explore how opposing muscles allow humans to bend and extend their arms.

Getting Started

1. Think about all the different joints of the body and the movements that these joints enable humans to make. Discuss your ideas with the group. Then list the ideas in your science notebook.

2. Share your ideas with the class.

Inquiry 19.1
Exploring Joints With Models

PROCEDURE

1. Have someone from your group pick up a tote tray. Note that it contains two kits. One kit has the materials to construct a model of a human arm, and the other kit contains materials to construct a spinal column. Each pair of students in your group will work with one kit. Your task is to assemble a working model from the materials in the kit.

2. With your partner, study the materials in your kit. Then decide on a plan for assembling the model (see Figure 19.1).

3. If the period ends before you finish assembling your model, take it apart and put the pieces back in the bag. Put the bag in the tote tray and return it to the designated area. Your teacher will return it to you at the beginning of the next period.

For you
  1 copy of Student
    Sheet 19.1:
    Exploring Joints

For your group
  1 tote tray
  1 arm model kit
  1 spinal column
    model kit
  1 roll of clear tape

Figure 19.1 *Assembling the model will take some planning.*

4. After you have assembled your model, share it with the other members of your group. Discuss the following points:

  • the names of the structures represented by the parts of the models (see Student Sheet 19.1)
  • the types of movement permitted at the joints of the model
  • how the movement at the joints is produced by the action of opposing muscles

5. Keep your model assembled until you have completed Student Sheet 19.1. Then take it apart and put the pieces back in the appropriate bag. Return the tote tray to the designated area.

6. Answer the following questions in your science notebook:

A. What are the steps involved in bending and extending your arm?

Because the spine has so many joints, humans can literally bend over backwards (with a little practice, of course).

B. What kinds of movements does the ball-and-socket joint, which is located at the shoulder, allow you to make?

C. What kinds of movements does the spinal column permit?

D. In addition to the places listed below, where can you find another of each of the following joints?

  • hinge joint: elbow and _____
  • ball-and-socket joint: shoulder and _____
  • pivot joint: neck and _____

REFLECTING ON WHAT YOU'VE DONE

1. With the class, discuss the following:

A. The joints you identified in your model and your answers to the questions in Step 6 of the Procedure.

B. Other joints of the body, including the pivot joint of the forearm.

C. Similarities and differences between the arm model and the chicken wing that you explored in Lesson 18.

How are the two structures comparable?

D. The opposing muscles of the upper part of the human arm (see Figure 19.2).

What are other examples of muscles that work in opposing pairs?

2. Revise the list you developed during "Getting Started."

Figure 19.2 *Extending and bending the arm*

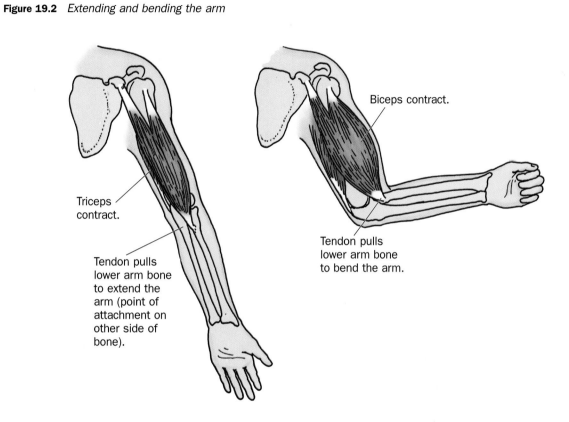

Triceps contract.

Tendon pulls lower arm bone to extend the arm (point of attachment on other side of bone).

Biceps contract.

Tendon pulls lower arm bone to bend the arm.

Major Joints of the Human Body

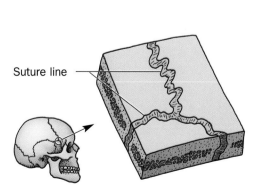

Suture line

Figure 19.3 *Fibrous joint of the skull*

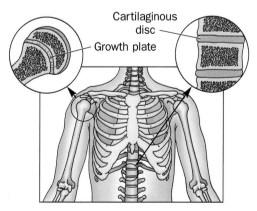

Cartilaginous disc

Growth plate

Figure 19.4 *Cartilaginous joints*

Major Joints of the Human Body (continued)

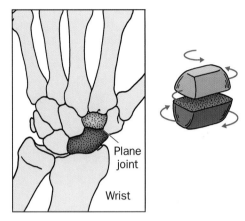

Figure 19.5 *Plane joint*

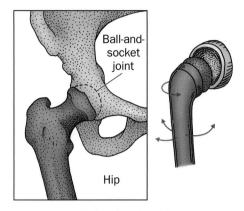

Figure 19.6 *Ball-and-socket joint*

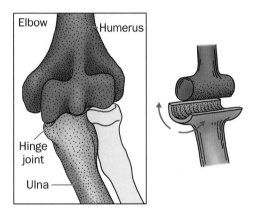

Figure 19.7 *Hinge joint*

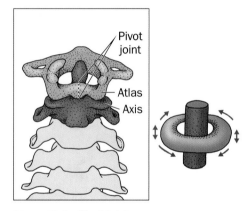

Figure 19.8 *Pivot joint*

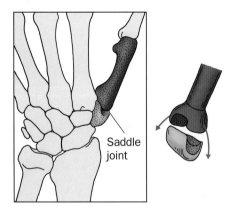

Figure 19.9 *Saddle joint*

WHAT KIND OF JOINT IS THIS?

Peppi and Bollo re-enter Joanne's body (this time, they slip in through a pore in her skin) and head for the shoulder. The spies are ready to take an inside look at the musculoskeletal system.

"This is a joint—a place where bones meet," says Peppi. "The joints are held together by those thick, elastic bands, called ligaments. Joints make it possible for the human body to bend, twist, and reach—to be flexible. Some people have even greater flexibility at their joints, because their ligaments stretch more than normal. These people are sometimes said to be 'double-jointed.'

"There are several kinds of joints, and each permits a human to move an arm, leg, finger, or whatever—in a different way. But since we're at the shoulder, let's stop and take a look. This is a ball-and-socket joint. Its design allows humans to move their arms in just about any direction."

"Kind of slippery in here," says Bollo as he touches the shiny white end of a bone.

The slippery synovial fluid lubricates the joints.

"Many joints need to be 'oiled' in order to work well, just like the moving parts of a machine. The joints are oiled by the body's homemade lubricant, which is called synovial fluid. Humans also have a special substance called cartilage at the ends of bones. The cartilage is softer and smoother than bone tissue. The cartilage, along with the synovial fluid, helps reduce the friction between bones."

Bollo sneaks down into the space where the shoulder bone and the upper arm bone meet. "Do you know what this joint reminds me of? A computer joystick!"

"Great comparison. Now let's move on down the arm," says

Peppi. "Watch how this joint works. What could you compare it with?"

Bollo watches. "It works like a door that can swing open and shut."

"Good comparison. This is the elbow. It is a hinge joint.

"We don't have time to visit all the types of joints. If we could, you'd see many varieties. For example, the head swivels on a pivot joint. The feet have gliding joints that help the body keep its balance. Wrists have special joints that let them move back and forth and side to side."

Teamwork

"Next," says Peppi, "we're going to explore how the bones, muscles, joints, and nerves work together. Stay there in the elbow. Look up. What do you see?"

"I see bundles of muscles," says Bollo. Some are right under the skin. Others cover the bone. The muscles are tapered at the ends. The muscles are connected to the bones by thin tissue . . .

"Tendons," says Peppi.

"I see one especially big muscle on the front of the arm and another big one at the back," Bollo continues.

"Right. The biceps muscles are just above the crook of the arm. The triceps muscles are on the opposite side of the arm—above the elbow. Keep your eye on those two muscles. I think a message is coming from command central."

Suddenly, everything seems to be happening at once. The biceps muscles contract and get fatter and shorter. The triceps, which are right behind it, relax

and lengthen. The lower part of the arm pulls upward.

Then everything happens in reverse. The biceps relax, and the triceps contract. The forearm is extended.

"Most skeletal muscles are like these two," says Peppi. "They work in pairs. They have to, because muscles can only do one thing— pull, or contract. They can't push."

"Can a muscle keep contracting forever? Does it ever need maintenance?" says Bollo.

"Muscles have endurance, but they cannot go on forever," says Peppi. "Muscles, like other body cells, need nutrients and oxygen from the blood. They convert the nutrients and oxygen into energy, store it, and tap into it when the brain tells them it's time to go to work. If muscles don't

get enough oxygen during hard exercise or work, they may start to burn or ache."

Back to the Bones

"That's the story on muscles and joints. But before we file our final report, we need to take a closer look at bones," says Peppi.

"We already know what bones do. They support the body and protect organs," says Bollo, checking his notes.

"That's what the skeleton does. But now we're going to take a look at a single bone," says Peppi.

"Why bother? Nothing's going on inside a bone, is there?" says Bollo.

"That's what people once thought. But they were wrong. Bone is living tissue. About 10 percent of bone tissue is replaced every year."

Peppi and Bollo make their way inside the bone. They cross

a membrane that is covered by veins and arteries. Next is a hard layer of dense bone. Deeper inside is a smaller area of lighter, spongy bone. At the center of the bone is soft tissue.

"Marrow," explains Peppi. "This is where the body manufactures blood cells."

"Look at them go!" says Bollo, taking a close look with his high-powered hand lens.

"Every second, more than 2 million red cells leave the marrow and enter circulation. White blood cells are made here, too."

"The human body is a hard worker," says Bollo. "Every organ and cell is doing its part—in fact, a lot of them, like bone, are doing many things at once."

"Once again you've learned your lesson, Bollo," says Peppi. "Now why don't you take a few moments to review the illustrations of the joints in our anatomy book. Then we'll be ready for another investigation that will give us a closer look at the muscles." □

Bones might be hard on the outside, but not necessarily in the middle, Bollo discovers.

20

Muscle Size and Strength

This tiny leaf-cutting ant can carry quite a load for its size. What's the connection between size and strength as far as humans are concerned? You're about to find out.

INTRODUCTION

You have probably seen photographs of body-builders. Those bulging muscles look so powerful! Or maybe you know some people who work out in a gym. They may compare notes about how much weight they can press.

How can you tell how strong muscles are? Is muscle size an indication of strength? In this lesson, you and your partner will measure the size of three muscle groups in your arm. Then you'll use a scale to measure the strength of those muscles. You will pool data from your group members and use that information to see whether muscle size is related to muscle strength. Let's see how strong your muscles really are!

OBJECTIVES FOR THIS LESSON

Measure and record the size and strength of three muscle groups in the arm.

Determine whether the size of three muscle groups in the arm is an indication of their strength.

Getting Started

MATERIALS FOR LESSON 20

For you
1 copy of Student Sheet 20.1: Muscle Size Versus Strength

For your group
1 bathroom scale
2 tape measures

1. Today, you are going to explore whether the strength of your arm muscles is related to their size. You will investigate three muscle groups: the biceps, triceps, and forearm muscles. Before you start, listen as a student volunteer reads the Introduction to this lesson from the Student Guide.

2. Look at the figures on pages 170–71 of the Student Guide as your teacher describes three muscle groups of the arm. Using a ranking system of 1 to 10 (with 10 being strongest and 1 being weakest), predict the relative strength of the three muscle groups. Record your predictions in your science notebook. Explain in writing why you ranked the muscles as you did.

3. Share your predictions with the class.

Inquiry 20.1
Investigating Muscle Size and Strength

PROCEDURE

1. Pick up your materials. During the first part of this inquiry, you will work in pairs to measure your muscles. You will then complete the inquiry with your group.

2. With the class, discuss the following questions:

A. What do you think the word "force" means?

B. How can you measure force?

C. How will you convert kilograms, which are the units of mass on the scale, to units of force?

3. Use the following procedure to measure the size of your partner's biceps, triceps, and forearm muscles.

A. Biceps. Have your partner sit at a desk with his or her arms positioned as seen in the illustration that follows. While your partner presses up against the bottom surface of the desktop with his or her palms, wrap the tape measure around your partner's left biceps in the place indicated in the figure. Record the left biceps measurement in the first column on Table 1 on *your partner's* student sheet. Repeat this procedure to measure and record data about your partner's right arm. Add the two measurements to obtain total biceps size. Record this on Table 1 on your partner's student sheet. If the desk raises up while your partner presses against it, have a third student sit on the desk to hold it steady.

Press your palms against the bottom of the desk while your partner measures each of your biceps.

B. Triceps. Have your partner position his or her arms as shown in the figure that follows. While your partner presses on the top of the desk with his or her palms, place the tape measure around his or her left arm in the same place that you did before. Record the triceps measurement on Table 1 on your partner's student sheet. Repeat this procedure to measure and record data about your partner's right triceps. Add the two measurements to obtain total triceps size. Record the total in Table 1.

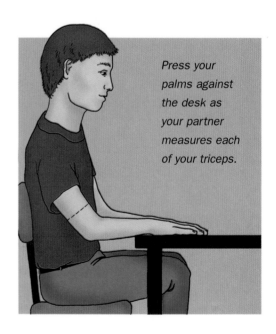

Press your palms against the desk as your partner measures each of your triceps.

C. Forearm muscles. Have your partner grasp the Student Guide with two hands in the position shown in the figure that follows. Wrap the tape measure around your partner's arm in the place shown. While your partner squeezes the Student Guide with both hands as hard as he or she can, measure the left forearm muscles and record that measurement in Table 1. Make sure that your partner does *not* squeeze the book with his or her thumbs. Repeat this procedure to measure and record data about your partner's right forearm muscles. Add the two measurements to obtain the total size of the forearm muscles. Record these data in Table 1.

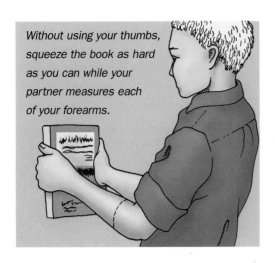

Without using your thumbs, squeeze the book as hard as you can while your partner measures each of your forearms.

4. Reverse roles with your partner and repeat Steps 3A through 3C.

5. Now that you have measured muscle size, it is time to focus on strength. Use the technique depicted in the three illustrations that follow as examples of how to measure the force applied to the scale by your biceps, triceps, and forearm muscles, respectively. When measuring your triceps, press with your hands, not with your arms. When measuring your forearm muscles, be sure to keep your thumbs away from the scale. After you have completed each measurement, record your results in the proper columns in Table 1.

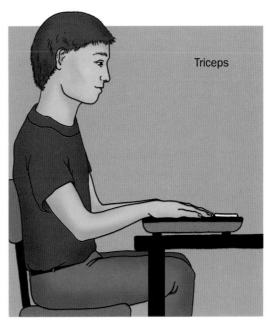

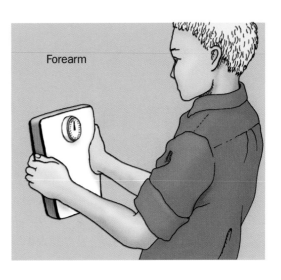

6. When everyone in your group has finished one set of tests, repeat the rotation if time permits. Calculate the average of your two sets of results and record the results on Table 2 on Student Sheet 20.1.

7. Return the scale and tape measure to the materials distribution center.

8. Complete Table 3 on Student Sheet 20.1.

9. Respond to the following questions in your science notebook:

   A. *In Table 3, you sorted your data on the basis of muscle size. Study the information in the table. What does the table indicate about the relationship between muscle size and strength in regard to the following muscles: biceps, triceps, forearm?*

   B. *On the basis of what you have observed in this inquiry, what is the relationship between muscle size and strength?*

   C. *Name at least one variable that you did not control in this inquiry. How could it have affected your data?*

REFLECTING ON WHAT YOU'VE DONE

1. Discuss the results of your inquiry and the answers to the questions in Step 9 of the Procedure.

2. Look at the predictions you made during "Getting Started." Were they accurate? If not, why?

3. The muscles you explored in this inquiry are skeletal muscles. Discuss the following questions with your class:

   A. *What two other types of muscles have you studied in this module?*

   B. *Where are they located?*

   C. *How do they differ from skeletal muscles?*

Strength, flexibility, and coordination are key elements in Bob Ojeda's pitching technique, which is shown in this stroboscopic view.

CORBIS/ROGER RESSMEYER

Anabolic Steroids—Not Worth the Risk

Bigger, stronger muscles. Improved athletic performance. That is what anabolic steroids promise. But what are steroids? And are these claims true?

Steroids are drugs made from hormones, the chemical messengers of the human body. The U.S. Food and Drug Administration has approved the use of selected anabolic steroids for treating specific types of anemia, some breast cancers, osteoporosis (a bone disease), and several other diseases. Anabolic steroids are available legally only by prescription. There is a "black market" in anabolic steroids, however, that is similar to the illegal drug trade.

Although they have benefits when used properly under a doctor's care, anabolic steroids have been misused. Professional athletes—for example, weight lifters and sprinters—who needed quick bursts of muscle power began to use them to increase their performance. Steroids gained notoriety during athletic competitions such as the Olympic Games. Since the mid-1980s, when a test was developed to detect the presence of steroids in the urine, 17 different steroids have been banned by the Olympic Committee. Olympic athletes are tested for steroids before they are allowed to compete.

There is no real evidence that steroids do enhance athletic performance. Nevertheless, illegal steroid use continues. What's more, it is not limited to adults. Doctors, parents, and coaches are concerned about a growing trend of illegal steroid use among young people. They are concerned because anabolic steroids are powerful drugs that, when misused, can pose serious health risks.

How Steroids Work

Anabolic steroids are derived from a hormone called testosterone. "Steroid" refers to the drug's chemical structure. "Anabolic" refers to the drug's effect on the body. Anabolic steroids cause the body to retain nitrogen, which is a key ingredient of amino acids. Amino acids are the building blocks of protein, which helps build muscle and other tissue. Testosterone can stimulate the formation of red blood cells, which enable the body to carry more oxygen.

Anabolic steroids have many harmful side effects. They can cause high blood pressure, infertility, yellowing of the skin, acne, and trembling, as well as emotional effects such as depression. Violent outbursts of anger, sometimes called "roid rage," can result from steroid use. Women who take steroids may develop facial hair or deepened voices. Their menstrual periods may stop. Men may become bald or develop breasts. Adolescents may

Anabolic steroids have many negative side effects.

develop these same side effects. In addition, they may stop growing prematurely.

Solving the Problem

Most young people are aware of the risks of steroids. A majority of them disapprove of athletes who take these drugs. What do you think teachers, coaches, parents, and doctors can do to get the message out? How would you explain the dangers of using anabolic steroids to one of your friends? What would most convince you never to use steroids? ☐

LESSON 21

Exploring Muscle Fatigue

When the U.S. women's team won the World Soccer Cup after a game that went into double overtime, their muscles were probably very fatigued. But for a few moments, they were still able to jump for joy!

INTRODUCTION

When a basketball player leaves the court during the third quarter of a close game, he looks totally exhausted. But after a few minutes on the bench, he's ready for action again. You've probably noticed the same thing yourself if you play soccer, tennis, or any active sport.

What causes muscle fatigue? Can you do anything to prevent it? These are important questions, especially if you like sports. And even if you don't, it's good to know how to keep your muscles in shape. In this lesson, you'll learn something about your own muscle endurance.

OBJECTIVES FOR THIS LESSON

Measure the rate at which muscles tire during exercise.

Plot data about muscle fatigue on a graph and analyze patterns.

Getting Started

1. While a student volunteer reads aloud the Introduction of this lesson, hold your arms out straight in front of you. Keep them there as long as you can.

2. In your science notebook, describe how your arms felt as time passed.

3. Discuss your observations with the class.

Inquiry 21.1
Working Against Fatigue

PROCEDURE

1. Pick up your materials. During this inquiry, you will work with your partner to determine how the muscles of your fingers, hand, and forearm tire with repeated exercise.

2. With your teacher, read the directions for performing the inquiry as described in Steps 3A through 3E of the Procedure. Before you begin the inquiry, design a data collection table in your science notebook.

MATERIALS FOR LESSON 21

For you
1 copy of Student Sheet 21.1: Plotting Muscle Fatigue Data

For your group
1 tote tray
2 test tube clamps
2 stopwatches
1 set of colored pencils

3. Decide who will be the timekeeper and who will exercise for the first round of the inquiry. Then do the following:

A. Hold the clamp in your *dominant* hand in the position shown here.

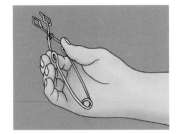

B. Squeeze the test tube clamp between your thumb and first two fingers of your hand until they meet; then, relax your grip until the clamp is back in its resting position. This is considered as one squeeze.

C. Using the stopwatch, your partner will record the number of squeezes you can do every 30 seconds for a total of 150 continuous seconds (five trials). Count the number of squeezes out loud.

D. At the end of each of the first four 30-second intervals, your partner will say "Time." Without pausing, continue to squeeze and begin counting again from 1 while your partner records the number of squeezes you made onto your table in your science notebook.

E. At the end of the fifth trial, your partner will say "Stop!" He or she will record the number of squeezes you made in the final 30 seconds. When this is done, switch roles and repeat Steps A through E.

4. When both you and your partner have completed five trials, share and record your data.

5. Return your materials to the designated area.

6. Graph the data for you and your partner on Student Sheet 21.1. Connect the dots using two different types of lines (for example, red lines for you and blue ones for your partner, or dashes for you and an unbroken line for your partner).

7. Answer the following questions in your science notebook:

A. What happened, in general, to the number of squeezes made per 30-second trial as the seconds passed?

B. How did your muscles feel at the beginning of the exercise? How did they feel when you continued to use them, even when they were tired?

C. On the basis of what you have learned about how your body releases and uses energy, explain why you think muscles tire.

Working in pairs to measure muscle fatigue

REFLECTING ON WHAT YOU'VE DONE

1. Discuss the results of your inquiry and reading.

2. What do you think you could do to increase your muscle endurance? Discuss your ideas with the class.

Repetitive Stress Injury:
Too Much of the Same Old Thing

They say that practice makes perfect, but sometimes doing the same thing over and over can cause problems.

One of these problems is repetitive stress injury. This kind of injury sometimes is associated with sports, and it has names like "tennis elbow" or "runner's knee." A repetitive stress injury that is becoming more common is caused by work and play. It's called carpal tunnel syndrome (CTS).

Carpal tunnels are narrow passageways inside the wrists. The carpal tunnels are crowded with nerves, blood vessels, and tendons that connect the muscles of your forearms with your hands and fingers. When you flex your forearm muscles to move your fingers, the tendons slide back and forth through the carpal tunnels.

Many people work at jobs—such as typing, cutting meat, or checking out groceries—that involve making the same movements with their hands hour after hour. This can cause pain and difficulty when they move their wrists and hands. The constant movement can cause the tendons in their wrists to swell.

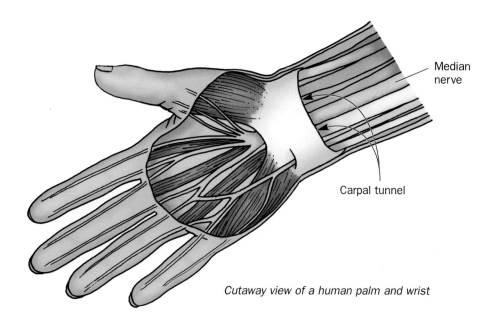

Cutaway view of a human palm and wrist

Median nerve

Carpal tunnel

This creates pressure on the nerves that run through the carpal tunnels. The result is a tingling or pain in the hand or arm. Sometimes the pain can extend to the neck and shoulders.

For some people, the discomfort becomes so bad that they can't work. Repetitive stress injuries, especially CTS, seem to be becoming more common. In 1985, 50,000 workers missed time from their jobs because of repetitive stress injuries; by 1995, the number had jumped to 300,000.

CTS doesn't cause problems only for adults. Young people can suffer from it, too. If you have a computer, you know it's a great way to do homework and a great place to play.

Sometimes students spend so much time typing reports, vaporizing aliens, and writing e-mail that they can develop CTS. They might have to wear a wrist brace or stay away from the keyboard, mouse, or joystick until their wrist heals.

What can you do to prevent CTS? Take a break. Stand up and stretch your hands, arms, and neck. When you move your mouse, use your whole arm, not just your wrist and hand. When you type, keep your hand and forearm in a straight line and make sure your wrists aren't resting on anything. Some keyboards help you to keep your hands in the right position—they are bent in the middle so you don't have to hold your wrists in an unnatural position. It's important to have the right kind of chair, too. Get one that supports your back.

If you're careful, you can spend plenty of time at the computer and not have to worry about anything more serious than getting the right answers for your homework! ☐

Some computer keyboards are especially designed to prevent CTS. Proper posture is important, too.

SPIES

COMMAND CENTRAL

Peppi and Bollo move upward from Joanne's shoulder toward her head.

"We're about to arrive in the brain. Time to stop and observe some important activities," says Peppi. "As far as the muscles are concerned, the brain is command central. Most muscles act only when the brain gives an order. When the brain says 'Sit,' certain muscles contract and pull. The bones and joints cooperate, and a human sits. More obedient than a trained dog!"

"How does that order reach the muscles?" asks Bollo.

"Think of the nerves as the messengers. They carry signals deep into the muscle cells. Inside the muscle cells are thousands of filaments. Normally, the filaments are slightly interlocked—like two combs with their teeth overlapping a little bit. When a nerve impulse reaches the muscle cells, the fibrils slide in between each other; this movement is called a 'muscle contraction.' The muscles go into action within a few thousandths of a second after getting their instructions."

"You said most muscles work this way. What about the others?"

"The muscles I've been describing are the skeletal muscles. They work when the nervous

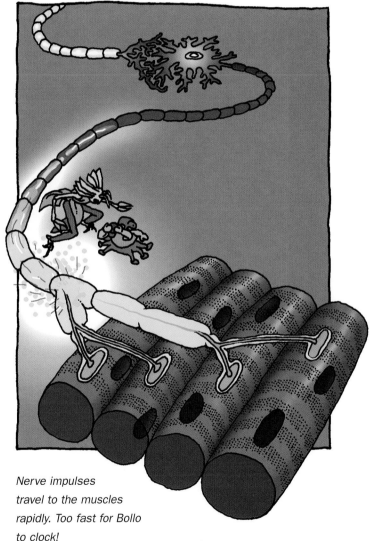

Nerve impulses travel to the muscles rapidly. Too fast for Bollo to clock!

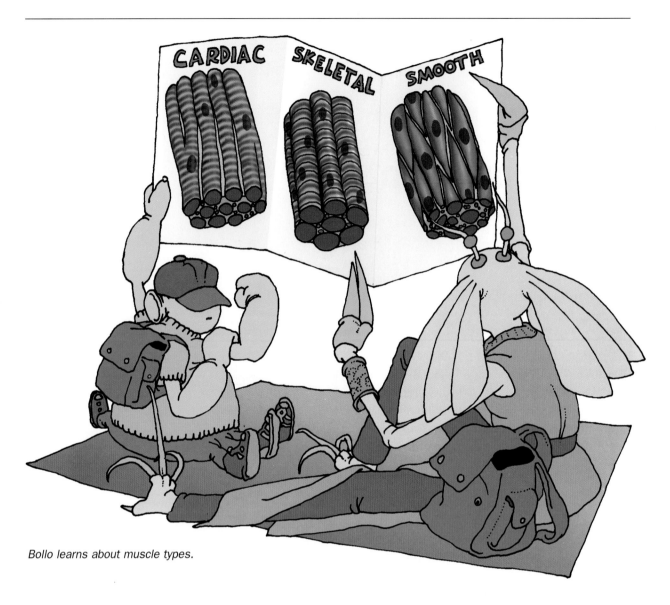

Bollo learns about muscle types.

system tells them to. But a few muscles are 'on automatic.' These include the muscles in the digestive system. As we saw during our trip through the digestive system, movement there is fairly constant."

"I bet the heart falls into that category,

too," says Bollo.

"You're right. The heart beats on its own . . ."

". . . because it gets a reminder from the heart's natural pacemaker, I know," says Bollo.

"Good for you!" says Peppi.

"There's still one

thing I don't understand," says Bollo. "Heart muscles work 24 hours a day. But when humans work or play hard, their skeletal muscles get really tired. Sometimes, they even cramp. What's the difference?"

"Different muscle types have different

designs. Heart muscles are designed especially for endurance," says Peppi. "Skeletal muscles, on the other hand, can tire. Some of them tire more easily than others."

"But why do they get tired?" asks Bollo.

"They run out of gas," says Peppi.

"Gas! There's no gasoline in muscles," says Bollo.

"You're right about that. But just like a car needs gasoline to keep it running, muscles need fuel, too. The fuel is the nutrients and oxygen that combine during cellular respiration. If the fuel supply gets low, the muscles begin to slow down."

"Can humans do anything to increase muscle endurance?" asks Bollo.

"Yes. Use them! When humans get plenty of exercise and use the same muscles again and again, their muscles become more resistant to fatigue," says Peppi. "Of course, they need to eat properly so that the muscles have the fuel they need to do their work. They also need to relax once in a while to give those muscles a break. Cardiac muscle, on the other hand, never takes a break."

"B-r-e-a-k! You said the magic word," says Bollo. "My muscles are ready for a rest."

"And so is Joanne," says Peppi. "Let's let her sleep in peace. It's time for us to pack up and head back home." ☐

Peppi and Bollo say good-bye.

22

The Body in Balance

An American engineer talks with some residents of Saudi Arabia in the city of Dhahran. Which of the people pictured is probably more comfortable in the desert heat? Why?

INTRODUCTION

Suppose you had to tell your heart to beat—or remind your eyes to blink. Can you imagine how different life would be? You would spend most of your time just trying to keep your body systems working.

Many functions of the human body are "on automatic"; others require conscious effort. In this lesson, you will discuss some of the things that your body does on its own. You also will perform an activity that will help you appreciate how hard it would be if you had to think about a function that your body handles automatically with ease. What is that function? Read on and find out!

OBJECTIVES FOR THIS LESSON

Discuss functions that the body does automatically.

Explore how to maintain a small amount of water at a constant temperature.

Apply observations from the inquiry to appreciate how the body maintains a constant internal temperature.

MAKING ADJUSTMENTS

When the temperature hits 35 °C and you can't head for the pool, what happens? You sweat. The evaporation of water from the surface of your skin is nature's cooling mechanism.

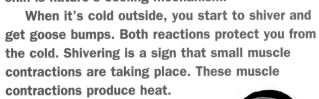

When it's cold outside, you start to shiver and get goose bumps. Both reactions protect you from the cold. Shivering is a sign that small muscle contractions are taking place. These muscle contractions produce heat. Goose bumps cause hair to stand on end and improve the body's insulation.

In both cases, your body has responded to a change in the external environment by making internal adjustments. Your body has maintained its homeostasis.

Humans are "programmed" to maintain homeostasis. Whenever something gets too far from normal, the body does its best to bring things back in line.

What are the advantages of homeostasis? How would human life be different if we did not have the ability to maintain homeostasis?

MATERIALS FOR LESSON 22

For you

1 copy of Student Sheet 22.1: Study Guide—Final Assessment
1 pair of safety goggles

For your group

1 tote tray
2 large test tubes
2 test tube clamps
2 beakers with ice water
2 candles
2 pie pans
2 thermometers
2 50-mL graduated cylinders
4 paper towels

Getting Started

1. Listen as one of your classmates reads the Introduction to this lesson. Then read "Making Adjustments," which follows the Introduction to this lesson in the Student Guide. Discuss the reading selection with the class.

2. With your group, develop a list of things your body does automatically. Record the list in your science notebook.

3. Discuss your responses with the class.

Inquiry 22.1
Maintaining a Balance

PROCEDURE

1. Review the Safety Tips.

> **SAFETY TIPS**
>
> Wear your safety goggles when you are near the candle flame. If your hair is long, keep it tied back.
>
> The candle and clamp may remain
>
> hot for a while after the flame has gone out. Leave the candle on the pie pan when you carry it back to the materials center.

2. With your teacher, review the Procedure for this inquiry. You will work in pairs. Pick up your materials.

3. Design a data table in your science notebook.

4. Now carry out the inquiry as follows:

A. Measure 15 mL of water in a graduated cylinder and pour it into the large test tube. Place the clamp on the test tube.

B. Place the thermometer in the test tube.

C. Hold the test tube so that the tip of the flame just touches the bottom of the test tube, as shown here. Keep the test tube in this position until the temperature of the water reaches 37 °C.

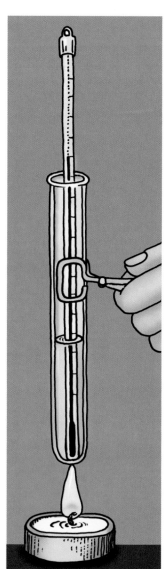

Hold the test tube just above the flame.

D. Maintain the temperature of the water at 37 °C by alternately holding the test tube above the flame and immersing it in the ice water. Move the tube between the two locations as quickly as possible. Record the temperature on your data table every minute for 10 minutes. The temperature at "0" minutes on your data table should be 37 °C.

E. Quickly touch the bottom of the test tube to a paper towel each time you remove the test tube from the ice water. Otherwise, water might drip from the test tube and extinguish the flame.

F. When 10 minutes have passed, record the temperature one final time. Dip the test tube into the ice water. Let the test tube cool while you finish recording your inquiry results.

5. Remove the test tube clamp and the thermometer. Pour the water into the sink or a container.

6. Return your equipment to the designated area.

7. Make sure you have completed the data table. Then answer the following questions in your science notebook:

A. Look at the variations in temperature on your data table. Did you have difficulty keeping the temperature at exactly 37 °C? If so, why?

B. What does the body do automatically to keep itself from getting too hot?

C. What does the body do automatically to keep itself from getting too cold?

REFLECTING ON WHAT YOU'VE DONE

1. Discuss the results of your inquiry with the class.

2. Discuss the following questions with the class:

A. What can you do voluntarily to avoid getting too hot? Too cold?

B. Why might you feel less comfortable on a humid day than on a dry day, even though the temperature was the same?

3. Take another look at the list your group made during "Getting Started." Revise it if necessary.

CORBIS/BETTMANN

Homeostasis comes from two Greek words that mean "staying the same." The concept was developed by Dr. Claude Bernard, a French physiologist of the 19th century.

Dr. Claude Bernard

CORBIS/VITTORIANO RASTELLI

How long would your nose be if it grew every time you told a lie—even a tiny one?

LIE DETECTORS: TRACKING REACTIONS

Do you remember Pinocchio? Whenever he told a lie, his nose got longer.

The modern technique of lie detection is based on a similar idea. The idea is that when you don't tell the truth, your body reacts.

Lie detection, or polygraphy, measures reactions that are harder to see than Pinocchio's expanding nose. It measures changes in blood pressure, pulse, and breathing rate, as well as galvanic skin response, or GSR. GSR is measured by placing electrodes on the hands. If the person's palms get sweaty, the electrodes measure changes in electrical resistance. Blood pressure and pulse rate are measured with a sphygmo-manometer similar to the one that doctors use.

More than a million polygraph tests are given in this country every year. Police use lie detectors in criminal investigations. Some employers use them to screen workers before hiring them. The use of lie detectors is controversial. Many states prohibit their use in employment applications, because the results are not always accurate or easy to interpret.

A Simple "Yes" or "No"

Pinocchio lied and lied, and his nose grew and grew. Then he started telling the truth, and each time he gave an honest answer, his nose got smaller. Polygraphy is more complicated than this. The polygraphers (people who administer lie detector tests) can't just hook up the machine and listen to what someone says. They have to ask specific questions that produce a "yes" or "no" answer.

Polygraphers ask two types of questions. One type is called a control question. It's a general question that doesn't have anything to do with the matter (often a crime) under investigation. A sample control question might be, "Do you live at 103 South Elm Street?"

The second type is the relevant question. It relates to the reason for the test. "Did you steal the purse of a woman at the corner of First and South Elm Streets on January 15?" is an example of a relevant question.

The theory behind polygraphy is that someone who is telling the truth will react more strongly to the control question than to the relevant ques-tion. The person who is lying, on the other hand, will react more strongly to the relevant question than to the control question. All these reactions are recorded on the polygrapher's testing devices.

It's Not Always Easy

Polygraphy can identify liars; however, it also produces a lot of "false positives." In other words, many innocent people test "guilty." This may be because they get anxious about taking the test. On the other hand, some liars are really smooth. They feel no emotion when they lie.

What's your opinion on lie detectors? Now tell the truth! ☐

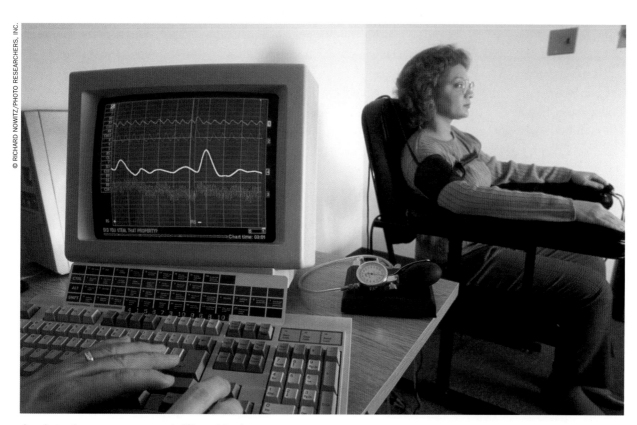

A polygraph measures several different body responses.

SPIES

BACK HOME

Peppi and Bollo emerge from their space capsule, pick up their gear from the trunk, and start to head home.

"See you tomorrow at headquarters, Bollo," says Peppi. "Remember, we've got to make our report at 9 a.m. sharp. I want you to take the lead on this."

"Can we get together early to review what we're going to say?" asks Bollo. "I learned so much that I don't know where to start."

"Sure. Let's meet around 8:30," says Peppi.

Talking It Over

"Just half an hour to get ready for our final report. Wow, am I nervous," says Bollo.

"Calm down," says Peppi. "You did a great job taking notes, asking questions, and exploring the human body systems. Now all you have to do is answer our leaders' questions: What makes the human body such an efficient organism? Why does it work so well?"

"Cooperation, for one thing," says Bollo. "The systems we explored all depend on one another. They work together. Here's a diagram I've made to illustrate how the systems interact."

"That's great. That diagram will be a good addition to our presentation," replies Peppi. "What else have you learned?"

"Specialization. Just think of the small intestine. The folds weren't enough. Humans also have villi—and even micro-villi. Every organ and

Home, sweet home!

Bollo prepares his report.

Ready to Take the Test

"I'm running out of ideas. What am I missing?" asks Bollo.

"The only one I can think of is that human body systems are dependable," says Peppi. "Sure, humans get ill sometimes. But if they eat the right foods, exercise regularly, and avoid alcohol, tobacco, and other unhealthy substances, they'll probably stay healthy—and live to be at least 80 years old. The average human life span is quite a bit longer than that of many other animals on planet Earth."

"You're right about that," says Bollo. "Okay. I've got my drawings and notes. I'm ready for the final assessment. Let's head up to the University of Saganova. And thanks for your help, Ms. Peppi. You've been a great teacher, and this has been quite a trip." □

tissue is needed to make the body work at peak efficiency."

"Keep going," urges Peppi.

"Things often come in pairs," says Bollo. "For example, humans have two lungs, two nostrils, two eyes and ears, and two arms and legs."

"What's the advantage there?" asks Peppi.

"I can think of a couple. One is for balance. But another might be for protection and survival. For example, if one eye got injured, humans would have another one—just in case. Or, if a loved one needed a kidney transplant, a relative could donate one kidney and live quite comfortably with a single kidney."

"Good thinking, Bollo!" says Peppi.

Final Assessment—
Human Body Systems

Boxers Julio Cesar Chavez (left) and Pernell Whitaker trade punches. For boxers and many other athletes, a quick reaction time is just as important as physical strength.

INTRODUCTION

What is reaction time? For a boxer, it's the time between the instant he sees a punch coming his way and the moment he tries to dodge it. For two people starting to walk across a busy intersection, it's the time between the instant they see a car barreling through the red light and the moment that they begin to jump back to the safety of the curb. In most cases, reaction time is so brief that it can be measured only in fractions of a second.

Even so, reaction times differ. Chances are, for example, that you would have a lot more trouble dodging those punches than a professional boxer would. That's normal. The boxer has trained for a long time to improve his reaction time. His ability to see the punch coming and move out of the way or block it in the shortest possible time is crucial to his career as well as to his health.

What is your reaction time? How fast do you react with the joystick when playing your favorite video game? In this lesson, which also serves as an assessment of the *Human Body Systems* module, you will learn something about your own

OBJECTIVES FOR THIS LESSON

Design and conduct an inquiry to explore the concept of reaction time.

Determine the effect of practice on reaction time.

Differentiate between voluntary and reflex actions.

Respond to a number of selected-response items.

reaction time. For Part A of the assessment, you and a partner will design and carry out an investigation to explore the effect of practice on the time it takes to catch a ruler when your partner drops it. For Part B, you will work individually to complete a set of selected-response questions.

In this lesson, in other words, you will be assessing your reaction time. At the same time your teacher will be assessing you!

Testing reaction time

MATERIALS FOR LESSON 23

For you

- 1 copy of Student Sheet 23.1a: Part A: Reaction Time Inquiry
- 1 copy of Student Sheet 23.1c: Part B: *Human Body Systems* Content Assessment—Selected-Response Items
- 1 copy of Student Sheet 23.1d: Part B: *Human Body Systems* Content Assessment—Selected-Response Items Answer Sheet

For you and your partner

- 1 copy of Student Sheet 23.1b: Reaction Time Conversion Table
- 1 ruler

PROCEDURE FOR PERIOD 1

1. Watch as your teacher and a student volunteer demonstrate how you will measure reaction time. Then look at Student Sheet 23.1b, which shows how to convert centimeters (cm) to seconds (sec).

2. Work with your partner to design an investigation to see whether you can improve the time in which you can catch a ruler when your partner drops it.

3. Using Student Sheet 23.1a, plan your investigation. When you have recorded answers to numbers 1 through 4, complete your investigation.

4. Consider using the Inquiry Checklist (SG page 63) and the Inquiry Scoring Rubric (SG page 63) to make sure you have done everything necessary for your inquiry.

PROCEDURE FOR PERIOD 2

1. Complete the selected-response items that appear on Student Sheet 23.1c. Mark your responses on Student Sheet 23.1d. Do *not* write on Student Sheet 23.1c.

2. When you have finished the assessment, turn in the test and the answer sheet to your teacher.

3. If you have time, begin reading "Reflex Actions: Ready or Not, Here They Come," at the end of this lesson.

REFLECTING ON WHAT YOU'VE DONE

1. Discuss with the class "Reflex Actions: Ready or Not, Here They Come."

2. Your teacher will return Student Sheet 23.1a and your answer sheet for Part B of the assessment. Follow along as your teacher reviews the questions and possible responses.

3. Think of human body systems that were not covered in this module. How could you find out more about them?

Reflex Actions: Ready or Not, Here They Come!

CORBIS/LESTER V. BERGMAN

Now do you understand why your parents tell you to cover your nose and mouth when you sneeze?

Something tickles your nose and . . . achoo! You don't think about doing it. You can't stop it. Sneezing is a reflex action.

Many of our actions are *voluntary*. In other words, we think before we act. Before you volunteer to speak up in class, for example, you ask yourself, "Do I really know the right answer?" If you do, you decide to raise your hand.

Reflex actions, by contrast, are involuntary. We do them without thinking. If someone sticks you in the arm with a pin, you jerk away automatically. Your body does this on its own— with no instructions from the brain.

Some reflex actions are a way for your body to protect itself. A sneeze, for example, is your body's way of getting rid of something that is irritating your nasal passages. Putting out your hands is a way to protect your body if you sense you are falling.

Reflexes begin when something stimulates your sensory

nerves. Take that pin, for instance. When it touches your skin, your sensory nerves send a message to your spinal cord. Your spinal cord then sends off messages in two directions. One message goes to the muscles of your arm. This is what makes you jerk your arm away. The other message goes to your brain, which interprets the pin prick as pain. Luckily for you, the message to your arm is transmitted faster than the message to your brain. That's why you jerk your arm away without even thinking about it!

Reflex actions can cause movement throughout the body. Consider that sneeze, for example. Muscles in your jaw move to open your mouth. Muscles in your chest and abdomen go into action to fill your lungs with air. Your tongue moves up against the roof of your mouth. Your

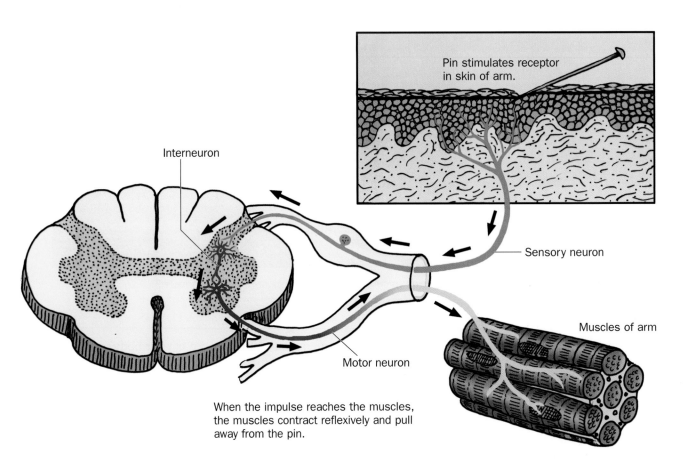

Interneuron

Pin stimulates receptor in skin of arm.

Sensory neuron

Muscles of arm

Motor neuron

When the impulse reaches the muscles, the muscles contract reflexively and pull away from the pin.

Anatomy of a reflex action

eyelids squeeze shut. Muscles in the back of your neck contract to tilt your head back. Then, your mouth closes and your diaphragm contracts rapidly. Air blasts through your nose at speeds up to 40 meters per second. Whatever caused the tickle goes flying out with the air.

So what's the difference between voluntary actions and reflex actions? Voluntary actions occur "on demand"— when the brain says so. Reflex actions occur instantly and automatically. If the choice is sink or swim, your brain doesn't have to say, "Swim!" Our bodies perform other actions automatically, too; sweating and breathing are two examples. These actions, however, do not occur as quickly as reflex actions do. ☐

Glossary

absorption: The process by which digested nutrients pass through the wall of the small intestine into the bloodstream.

active transport: The process by which materials, using energy supplied by the cell, are moved across a membrane. *See also* **passive transport**.

alveoli: Tiny sacs in the lungs through which the exchange of oxygen and carbon dioxide takes place. The singular form of alveoli is "alveolus."

amino acids: The building blocks of proteins.

amylase: A digestive enzyme that breaks down starch.

antibiotic: A medication that weakens or kills bacteria.

antigen: A substance that is recognized as foreign by the immune system and that causes the immune system to produce a specific antibody to it.

ATP (adenosine triphosphate): An organic chemical in which energy is stored and from which energy is released to meet the body's needs.

atrium: One of the two thin-walled upper chambers of the heart. The plural of atrium is "atria."

ball-and-socket joint: A joint that permits movements in all directions. Ball-and-socket joints are found at the shoulders and hips.

Benedict's solution: A chemical indicator that, when added to a solution and heated, changes from blue to light green to red in the presence of increasing concentrations of sugar.

blood pressure: The force exerted by blood against vessel walls.

body system: A group of organs that work together to perform a specific function. For example, the organs of the digestive system process food and prepare it for delivery to the body's cells.

breathing: The mechanical process of moving air into and out of the lungs; also called ventilation.

bromthymol blue: A chemical indicator that changes from blue to light green to yellow in the presence of increasing concentrations of carbon dioxide in solution.

calorie: A unit of heat energy. Spelled with a lowercase "c," the word "calorie" describes the amount of energy required to raise the temperature of 1 gram of water 1 degree Celsius. Spelled with an uppercase "C," the word "Calorie" describes the amount of energy required to raise the temperature of 1000 grams of water 1 degree Celsius.

carbohydrates: One of the three basic food types. May be in the form of starch, sugar, or fiber. Found in cereals, breads, and vegetables.

cartilage: Firm, smooth tissue at the ends of bones. Provides a smooth surface between bones. Also present in areas such as the nose, ears, and voice box.

cell: The smallest unit of an organism that can carry out the basic functions of life.

cellular respiration: The process by which glucose combines with oxygen to produce energy, carbon dioxide, and water.

cholesterol: A lipid found in animal fat and most animal tissue. Synthesized by the liver; also part of many foods.

chyme: A pulpy mixture of food and gastric juices. Produced in the stomach, from which it passes into the small intestine.

cilia: Tiny hairlike extensions of cells that line the respiratory tract. The cilia move in a wavelike fashion to help eliminate dust and germs from the body.

closed circulatory system: A transport system in which the fluid is confined within vessels. Humans have closed circulatory systems. *See also* **open circulatory system**.

combustion: A rapid form of oxidation that releases heat and, in many cases, light.
contagious: Capable of being transmitted from person to person.

diffusion: The process by which molecules move from places where they are more concentrated to places where they are less concentrated. A form of passive transport.
disease: Any disorder or malfunction of the body or a part of the body. May be caused by internal factors (for example, heredity, which causes certain forms of heart disease) or external factors (for example, bacteria, which cause tuberculosis).
duodenum: The first 25 centimeters of the small intestine; site where most chemical digestion occurs.

endocrine system: A network of glands and cells that secretes chemical messengers (hormones) into the bloodstream or lymph.
energy: The ability to perform work; may be stored in cells as fat or glycogen as well as in ATP.
enzyme: A protein that is capable of speeding up a chemical reaction.
extensor: A muscle that extends, or straightens, a body part by increasing the angle at a joint.

fats: (1) One of the three basic food types; found in oils and some dairy products.
(2) Tissue that provides a cushion for various body parts, insulates the body, and stores energy in a concentrated form.
feces: Solid wastes in the large intestine that are expelled from the body during bowel movements (defecation).

fiber: An indigestible carbohydrate such as cellulose that stimulates peristalsis in the intestine.
flexor: A muscle that bends a body part by decreasing the angle at a joint.

gastric juice: A liquid that includes hydrochloric acid and pepsin and that is responsible for the chemical digestion of protein in the stomach.

heartburn: A painful sensation in the lower esophagus or upper stomach; sometimes caused by excess stomach acid.
heart murmur: An abnormal heart sound such as that caused by the flow of blood through a damaged valve.
heart rate: The number of times the heart beats in a given unit of time (usually one minute).
hinge joint: A joint that permits a back-and-forth movement similar to that of a door; the elbow is an example.
hormone: A chemical messenger produced by the endocrine system and transported through the bloodstream or lymph to certain target sites.
hydrochloric acid: A component of gastric juice that helps create the environment that pepsin needs to break down protein in the stomach.

immunity: State of being resistant to a disease-causing agent such as the polio virus or the bacterium that causes tuberculosis.
indicator: A substance that changes in some way to indicate the presence of another substance. Examples include Benedict's solution and Lugol solution.
infectious: Describes a microorganism that is capable of passing from one person to another.

integumentary system: The system that provides protective coverage for the body; the skin.

joint: A place where bones meet.

ligament: Tough, fibrous tissue that connects one bone to another bone.

Lugol solution: A yellow-brown indicator that turns blue-black when it comes into contact with starch.

macrophage: A protective cell in the blood, lymph, and connective tissue that engulfs and destroys bacteria and other foreign substances.

mucus: A thick, sticky substance that lines and protects the inner walls of the digestive organs. Facilitates the passage of food through the digestive tract and helps protect the walls of the digestive tract from being digested.

nervous system: The body's "control" system; initiates muscle contractions and glandular secretions.

open circulatory system: A transport system in which the blood is not confined within vessels. Many insects have open circulatory systems. *See also* **closed circulatory system**.

opposing muscles: Muscles that work against each other at a particular joint so that the joint can move. For example, the biceps and triceps enable humans to bend and extend the arm.

organ: A group of different tissues that work together to perform a specific function. Examples include the heart, liver, and brain.

organism: A complete living thing. Members of the animal kingdom, such as humans, dogs, and cats, are organisms. Plants are organisms, as are bacteria and fungi.

oxidation: The process by which substances combine with oxygen.

pacemaker: A group of specialized cells in the right atrium of the heart that establish the basic rhythm of the heartbeat.

passive transport: A process in which substances pass through a cell membrane from a place where they are more concentrated to a place where they are less concentrated, without using any energy from the cell. *See also* **active transport**.

pathogen: A disease-causing agent; for example, a tuberculosis bacterium or a polio virus.

pepsin: An enzyme in the stomach that breaks down protein.

peristalsis: Regular muscular contractions that move food through the digestive tract.

pivot joint: A joint that permits movement of one bone around its own long axis or around the axis of another bone; for example, the atlas joint at the neck.

plaque: The buildup of materials on the inner wall of a blood vessel.

plasma: The liquid part of the blood; makes up about 55 percent of the blood.

platelets: Cell fragments in the blood that aid in clotting.

proteins: One of the three basic food types; needed for building and repair of tissue in the body. Found in beef, egg whites, nuts, and pork. All enzymes are proteins.

pulmonary circulation: The vessels that transport blood from the heart to the lungs and back to the heart.

pulse: The rhythmic expansion and recoil of arteries; initiated by the contractions of the ventricles of the heart.

residual volume: The amount of air that remains in the lungs after a person exhales as forcefully as he or she can.

Rh factor: A group of antigens on red blood cells; named after the Rhesus monkey in which it was first found; blood that has this group of antigens is called Rh-positive; blood that does not have these antigens is called Rh-negative.

rubric: An established set of guidelines for assessing work.

saliva: Watery substance secreted by three pairs of glands around the mouth. Helps moisten and soften food for swallowing. Contains an enzyme called amylase that begins the digestion of starch.

sphincter: A ring of muscle that aids in the one-way passage of food through the digestive tract.

spirometer: A device for measuring lung volume.

surface area: The part of an object that makes direct contact with its environment.

system: A number of parts that work together as a whole.

systemic circulation: The blood vessels that carry blood from the heart to the body and back to the heart. *See also* **pulmonary circulation**.

tendons: Tough, fibrous tissue that attach muscle to bone.

tissue: A group of similar cells that work together to carry out a specific function. The function of muscle tissue, for example, is to contract.

total lung capacity: The amount of air the lungs can hold after taking as deep a breath as possible (the sum of vital capacity and residual volume).

ulcer: An open sore or lesion in the skin or mucous membrane.

valve: A structure in the heart and some veins that prevents the blood from flowing backward.

ventilation: The movement of air into and out of the lungs.

ventricle: One of the two thick-walled lower chambers of the heart; the pumping part of the heart.

villi: Microscopic, fingerlike projections that line the inner wall of the small intestine and increase the surface area available for absorption of nutrients.

vital capacity: The total amount of air that a person can exhale after taking as deep a breath as possible.

Index

Photo Credits